MATH ADVENTURES
A Key to Academic Math Advancement

GRADE 1

_____

_____

_____

Author: Ace Academic Publishing

Ace Academic Publishing is a leading supplemental educational workbook publisher for grades K-12. At Ace Academic Publishing, we realize the importance of imparting analytical and critical thinking skills during the early ages of childhood and hence our books include materials that require multiple levels of analysis and encourage the students to think outside the box.

The materials for our books are written by award winning teachers with several years of teaching experience. All our books are aligned with the state standards and are widely used by many schools throughout the country.

Prepaze is a sister company of Ace Academic Publishing. Intrigued by the unending possibilities of the internet and its role in education, Prepaze was created to spread the knowledge and learning across all corners of the world through an online platform. We equip ourselves with state-of-the-art technologies so that knowledge reaches the students through the quickest and the most effective channels.

For inquiries and bulk orders, contact Ace Academic Publishing at the following address:
Ace Academic Publishing
3031 Village Market Place,
Morrisville, NC 27560, USA
www.aceacademicpublishing.com

This book contains copyright protected material. The purchase of this material entitles the buyer to use this material for personal and classroom use only. Reproducing the content for commercial use is strictly prohibited. Contact us to learn about options to use it for an entire school district or other commercial use.

ISBN: 978-1-949383-55-3
© Ace Academic Publishing, 2023

Introduction

About the Book

Welcome to "**Math Adventures - A Key to Academic Math Advancement**"! This workbook is specifically designed to align with the school curriculum and help students improve their analytical and logical thinking skills. With over **750 questions and several word problems**, this book aims to cover all the required syllabus for students in Grade 1.

Our workbook is an excellent resource for end-of-the-year state tests given by schools, as well as a great review book during the summer. Whether you are looking to improve your math skills or simply keep them sharp, "**Math Adventures**" provides a comprehensive and challenging set of problems to help you achieve your goals.

Our authors have extensive experience in teaching and developing math curricula for students at all levels. **They have carefully crafted each problem to challenge students and help them develop key problem-solving and critical thinking skills.** The book covers a wide range of topics, including arithmetic, algebra, geometry, and data analysis, providing students with a well-rounded education in math.

We believe that with practice, anyone can master math. "**Math Adventures**" is designed to help students build confidence in their abilities and develop a love for the subject. With clear explanations, helpful hints, and detailed solutions, this book is an excellent tool for anyone looking to improve their math skills.

Thank you for choosing "**Math Adventures - A Key to Academic Math Advancement**". We hope that you find it useful and enjoyable!

Common Core Math Workbooks

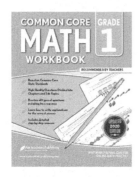   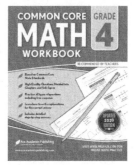

   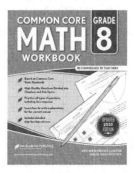

Common Core English Workbooks

The One Big Book Workbooks

Math Adventures Workbooks

Early Learning Workbooks

 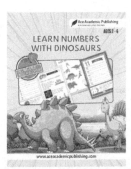  

TABLE OF CONTENTS

CHAPTER 1 ADDITION WITHIN 20

1.1 Addition Basics — 12
1.2 Finding Missing Addend — 19
1.3 Addition With Counting on and Making Ten — 25
1.4 Addition Word Problems — 31
1.5 Chapter Review — 35

CHAPTER 2 SUBTRACTION WITHIN 20

2.1 Subtraction Basics — 44
2.2 Subtraction by Decomposing a Number into Ten — 51
2.3 Subtraction Word Problems — 57
2.4 Chapter Review — 62

CHAPTER 3 PROPERTIES OF ADDITIVE AND SUBTRACTIVE OPERATIONS AND EQUATIONS

3.1 Commutative Property and Associative Property of Addition — 71
3.2 Three Number Addition or Subtraction — 78
3.3 Add or Subtract Within 100 Using Properties of Operations — 84
3.4 Finding Two Values Are Equal — 90
3.5 Chapter Review — 95

CHAPTER 4 PLACE VALUE

4.1 Count and Compare Numbers Within 120 — 104
4.2 Tens and Ones — 110
4.3 Finding Sums and Differences — 117
4.4 Chapter Review — 123

TABLE OF CONTENTS

CHAPTER 5 MEASUREMENT

5.1 Compare and Measure Lengths — 132
5.2 Clock and Time — 139
5.3 Chapter Review — 145

CHAPTER 6 REPRESENT AND INTERPRET DATA

6.1 Organize and Represent Data in a Chart — 154
6.2 Organize and Represent Data in a Table — 161
6.3 Interpret Data with Up to Four Categories — 169
6.4 Chapter Review — 178

CHAPTER 7 GEOMETRY

7.1 Geometric Properties - 2D and 3D Shapes — 189
7.2 Identify and Draw 2D and 3D Shapes — 196
7.3 Create Composite Figures in Two or Three Dimensions — 202
7.4 Chapter Review — 208

CHAPTER 8 PARTITION SHAPES

8.1 Divide Circles and Rectangles into Two or Four Equal Parts — 216
8.2 Describe Shares as Halves, Fourths, and Quarters — 222
8.3 Identify Equal and Unequal Shares — 228
8.4 Chapter Review — 235

COMPREHENSIVE ASSESSMENT I — 243
COMPREHENSIVE ASSESSMENT II — 255
ANSWERS AND EXPLANATIONS — 268

CHAPTER 1

ADDITION WITHIN 20

DOT-TO-DOT DINOSAUR

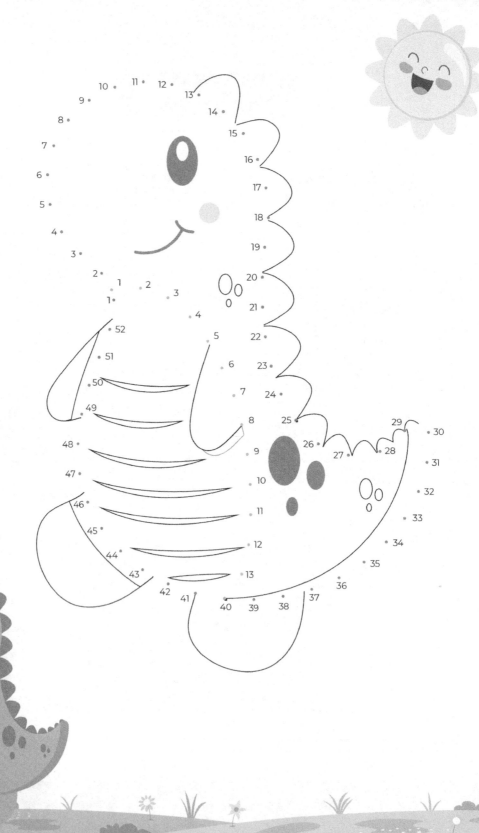

ADDITION WITHIN 20

ADDITION BASICS

Addition is nothing but combining two or more numbers or objects to find a total. It is one of the fundamental concepts in mathematics.

In the addition process, the numbers being added are called "addends," and the result is known as the "sum."

Example:

4 + 2 = 6

ADDITION WITHIN 20

1.1 Addition Basics

1 Count and write the total number of objects. Find the sum.

_____ balls + _____ balls = _____ balls in total.

2 Count and write the total number of objects. Find the sum.

_____ flowers + _____ flowers = _____ flowers in total.

3 Find the sum.

Ⓐ 8   Ⓑ 11   Ⓒ 7   Ⓓ 9

4 Count the dot on the dice and add them.

_____ + _____ = _____ dots in all.

ADDITION WITHIN 20

Addition Basics 1.1

Hint: When you add, Count up from the larger number.

Example: 2 + 8

> 2 + 8 is Harder: "2 ... 3, 4, 5, 6, 7, 8 , 9, 10"

> 8 + 2 is Easier: "8 ... 9, 10"

> So add 8 + 2 instead. You get the same answer.

5 Count the cones. Then fill in the correct answer.

- A) 8
- B) 11
- C) 7
- D) 9

6 Use the number line to find the answer: 5 + 4 = _____.

- A) 7
- B) 11
- C) 13
- D) 9

7 Count and write the total number of objects.

_____ + _____ = _____.

13

ADDITION WITHIN 20

 1.1   Addition Basics

8 Count the dots on the dice.
Then find the total number of dots.

_____ + _____ = _____ dots in all.

9 Count and write the total number of fish.

_____ + _____ = _____ in total

10 Add the two numbers to find the sum.

9 + 4 = _____.

Sum means total!

11 Use a number line to find the answer. 2 + 10 = _____.

0 1 2 3 4 5 6 7 8 9 10 11 12 13

 7   11   13   D) 12

14

ADDITION WITHIN 20

Addition Basics 1.1

QUICK TIP

You can use a number line to help you add 7 + 3 + 5 = ?

> Step 1: Start at 7.
> Step 2: Add 3.
> Step 3: Add 5.
> The answer is 15.

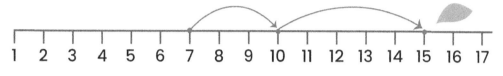

12 Count and write the total number of insects.

_____ + _____ = _____ in all.

13 Count the cubes. Then fill in the correct answer.

Ⓐ 6   Ⓑ 8
Ⓒ 9   Ⓓ 5

ADDITION WITHIN 20

1.1 Addition Basics

14 Complete the number sentence.

_____ + _____ = _____ in total.

15 Add the numbers. Then write the sum. 15 + 3 = _____.

16 Tom had 9 toys. He got 6 more toys at the store. How many toys does he have now in total?

(A) 7 (B) 11 (C) 13 (D) 15

17 Use a number line to find the answer.

7 + 3 = _____ .

0 1 2 3 4 5 6 7 8 9 10 11 12 13

ADDITION WITHIN 20

Addition Basics 1.1

18 Alice bought 6 candies. Her mother gave her 5 more candies. How many candies does Alice have now? Fill in the correct answer.

- A) 17
- B) 11
- C) 12
- D) 15

19 Find the total number of objects.

- A) 17
- B) 11
- C) 13
- D) 15

20 Add the two numbers to find the sum. Fill in the correct answer.

- A) 16
- B) 18
- C) 13
- D) 15

Next Section: Finding Missing Addend

ADDITION WITHIN 20

FINDING MISSING ADDEND

An addend refers to a number or quantity that is added to another number or quantity to get a sum. For example, in the equation 3 + 4 = 7, 3 and 4 are both addends, and their sum is 7.

$$4 + 3 = 7$$
Addend   Addend   Sum

Finding missing addend means determining what number is missing from an equation to make it correct.

Example:

Use the picture to find the missing addend

Explanation:

5 + 3 = 8, Missing number is 3.

ADDITION WITHIN 20

Finding Missing Addend 1.2

1 Which number goes in the blank to make the equation true?

5 + ☐ = 10

Equation means the same thing as number sentence!

2 Use the picture to help you find the missing addend.

12 = ☐ + 3

3 Use the objects to find the missing addend.

☐ + 3 = 12

4 Which number should be placed in the blank to ensure that the addition is correct?

7 + 7 = ☐   Ⓐ 14   Ⓑ 13   Ⓒ 10   Ⓓ 15

ADDITION WITHIN 20

1.2 Finding Missing Addend

5 Use the picture and find the missing addend.

5 + ☐ = 11

6 Find the missing number.

☐ = 3 + 2

7 Which number should be placed in the blank to ensure that the addition is correct?

9 + ☐ = 11   Ⓐ 4   Ⓑ 2   Ⓒ 5   Ⓓ 3

8 Use the picture to find the missing addend.

7 + ____ = 9

ADDITION WITHIN 20

Finding Missing Addend 1.2

9 Find the missing addend.

A) 7 B) 5
C) 9 D) 8

10 Find the missing addend.

A) 6 B) 8
C) 2 D) 3

11 Find the missing addend.

4 + ☐ = 8

A) 6 B) 8
C) 10 D) 4

12 Use the number line to find the missing addend.

5 + ☐ = 8

ADDITION WITHIN 20

1.2 Finding Missing Addend

13 Use the picture to find the missing addend.

 + ☐ = 8   Ⓐ 2   Ⓑ 4   Ⓒ 3   Ⓓ 1

14 Find the missing addend using the picture.

   ☐ + 2 = 5

15 Which number should be placed in the blank to ensure that the addition is correct?

☐ + 3 = 10   Ⓐ 3   Ⓑ 7   Ⓒ 2   Ⓓ 4

16 Use the picture to find the missing addend.

   12 = ☐ + 5

ADDITION WITHIN 20

Finding Missing Addend 1.2

17 Which number should be placed in the blank to ensure that the addition is correct? _____ + 7 = 18

- (A) 11
- (B) 12
- (C) 15
- (D) 13

18 Choose the correct answer to find the missing number.

- (A) 1
- (B) 2
- (C) 3
- (D) 4

19 Find the missing number using the picture.

- (A) 4
- (B) 6
- (C) 5
- (D) 3

20 Which number should be placed in the blank to ensure that the addition is correct? 9 + _____ = 14

- (A) 8
- (B) 7
- (C) 6
- (D) 5

Next Section: Addition With Counting on and Making Ten

ADDITION WITHIN 20

ADDITION WITH COUNTING ON AND MAKING TEN

Addition with counting on and making ten refers to a mathematical strategy used to find the sum of two numbers.

Counting-on method, the larger number is stated first, and then we count on from there by adding the smaller number.

For example, to add 6 and 3 using counting on, we start with 6 and count on 3 more numbers: 7, 8, 9. The sum of 6 and 3 is 9.

Making ten involves finding the missing addend to make a given number add up to ten. In the context of the word problem,

For example, Emma initially has 9 cupcakes and wants to reach a total of 10 cupcakes. The missing addend represents the number of cupcakes her friend gives her to reach the desired total (1).

This type of making ten word problem allows students to practice addition by finding the missing addend, developing their understanding of number relationships, and improving their mental math skills.

ADDITION WITHIN 20

Addition With Counting on and Making Ten **1.3**

1 Ben has 7 apple stickers and 8 orange stickers. How many stickers does he have in total using the counting-on method?

(A) 14   (B) 15   (C) 12   (D) 11

2 Alice has 6 kiwis and 3 dragon fruits. How many fruits does she have in total using the counting-on method?

(A) 9   (B) 12   (C) 16   (D) 10

3 Ava has 7 red marbles and wants to have a total of 10 marbles for her party. How many more marbles does Ava need to make a total of 10?"

_____

4 Emily has 6 candies, and her friend gives her some more. Now she has 10 candies in total. How many candies did Emily's friend give her?"

ADDITION WITHIN 20

1.3 Addition With Counting on and Making Ten

5. Tom has 6 apples and 5 oranges. How many fruits does he have in total using the counting on method?

(A) 11 (B) 12 (C) 13 (D) 14

6. Which number goes in the blank? Write 10+ strategies.
4 + 8 = 10 + _____.

7. David has 9 balls and 4 bats. How many toys does he have in total using counting?

(A) 18 (B) 13 (C) 15 (D) 11

8. Bella has 4 pens. How many more pens does she need to have a collection of ten pens?

_____

9. The equation "9+2"= ? can simplified and write 10+ strategies as

(A) 10 + 1 = ? (B) 10 + 2 = ? (C) 10 + 3 = ? (D) 10 + 0 = ?

ADDITION WITHIN 20

Addition With Counting on and Making Ten — 1.3

10. Which number goes in the blank?

$$10 + _____ = 9+7$$

(A) 5   (B) 6   (C) 3   (D) 7

11. Sam has 5 green beads and 7 yellow beads. How many beads does he have? 10 + _____ = _____ .

12. The equation "8+5"= ? can simplified and write 10+ strategies as

(A) 10 + 1 = ?   (B) 10 + 2 = ?   (C) 10 + 3 = ?   (D) 10 + 0 = ?

13. Which number goes in the blank?

$$10 + _____ = 9 + 9$$

(A) 6   (B) 2   (C) 9   (D) 8

14. Which number goes in the blank?

$$10 + _____ = 5 + 9$$

(A) 1   (B) 2   (C) 3   (D) 4

ADDITION WITHIN 20

1.3 Addition With Counting on and Making Ten

15 Jesrry has 8 toy cars, and he wants to make a set of ten toy cars. How many additional toy cars does Jerry, need?

_____

16 Dhoni has 3 red kites and 8 green kites. How many kites does he have in total using the counting on method?

A) 11   B) 10   C) 14   D) 16

17 Reethu has 7 butterfly stickers and, 5 lion stickers. How many stickers does he have in total? Write the related 10+ facts.

10 + _____ = _____ .

18 Solve the number sentence and write the 10+facts.

9 + 6 = _____ ,

10 + _____ = _____ .

ADDITION WITHIN 20

Addition With Counting on and Making Ten — 1.3

19 Solve the number sentence and write the 10+facts.

$$7 + 4 = \underline{\hspace{2cm}},$$
$$10 + \underline{\hspace{2cm}} = \underline{\hspace{2cm}}.$$

20 George has 5 crayons and wants to have a total of ten crayons. How many additional crayons does George need?

_____

Next Section: Addition Word Problems

ADDITION WITHIN 20

ADDITION WORD PROBLEMS

Addition is nothing but combining two or more numbers or objects to find a total. It is one of the fundamental concepts in mathematics.

The word problems in addition basically comes under two categories.

Category 1: Put together two or more collections of objects

Category 2: Increasing the number of objects involved in the situation.

Example:

Max has 8 strawberry candies and 6 chocolate candies. What is the total number of candies Max having?

The total number of candies Max has is 8 + 6 = 14.

ADDITION WITHIN 20

Addition Word Problems 1.4

1 Antony has 4 parrots and 8 love birds. How many birds does Antony pet to have?

(A) 12   (B) 16   (C) 18   (D) 20

2 Jack counted 6 cars in the parking lot. He saw 5 more on the road. How many cars did Jack see in all?

(A) 13   (B) 15   (C) 11   (D) 19

3 Mike went to the shop and bought one sandwich at 6 dollars and one Pepsi at 8 dollars. How much did he spend in all?

(A) 11   (B) 14   (C) 17   (D) 19

4 Jenny picked 5 lotus flowers and 8 lilies from the pond. How many flowers does she have in all?

(A) 10   (B) 17   (C) 15   (D) 13

5 At the grocery store, my mom bought 5 mangoes, 6 papayas, and 3 oranges. How many pieces of fruits did my mom buy?

(A) 14   (B) 16   (C) 11   (D) 19

ADDITION WITHIN 20

1.4 Addition Word Problems

6 Jessy has 8 cupcakes. Her friend gives her 5 more cupcakes. How many cupcakes does she have now?

- A) 16
- B) 13
- C) 11
- D) 19

7 Mellisa has 6 cows and 6 goats on the farm. Two more cows are added. How many animals does she have on the farm?

- A) 16
- B) 19
- C) 11
- D) 14

8 Jonny has 9 crayons and Steffi has 6 crayons. How many crayons do they have in total?

- A) 15
- B) 19
- C) 11
- D) 14

9 Teena has 5 green beads, 7 blue beads and 6 black beads. How many beads does she have in total?

- A) 15
- B) 18
- C) 11
- D) 14

10 The hen lays 6 eggs. Then, she lays 4 more. How many did she lay in all?

- A) 15
- B) 18
- C) 11
- D) 10

11 Angel has 3 mangoes. Mary has 5 more mangoes than Angel. How many mangoes does Mary have?

- A) 8
- B) 6
- C) 11
- D) 10

ADDITION WITHIN 20

Addition Word Problems 1.4

12. Blessy has badminton practice 11 days this month and 8 days next month. How many days of practice does she have in total?

A) 19   B) 6   C) 11   D) 10

13. Jamie ordered 5 burgers, 4 pizzas and 3 sodas. How many items did Jamie order altogether?

A) 14   B) 15   C) 12   D) 13

14. Tom found 8 seashells and Jerry found 2 seashells at the beach. What is the sum of their seashells?

A) 10   B) 11   C) 9   D) 12

15. William has 5 video games. Chris gives him 2 more video games. How many video games does William have altogether?

A) 2   B) 8   C) 9   D) 7

16. Max bought 8 candles. His mom gave him 5 more candles. How many candles does he have now?

A) 13   B) 11   C) 8   D) 17

ADDITION WITHIN 20

1.4 Addition Word Problems

17 Jack has 12 balls. His friend gave him 7 more balls. How many more balls does he have now?

- (A) 13
- (B) 19
- (C) 15
- (D) 17

18 There are 7 red colored eggs, 3 yellow colored eggs, and 2 blue colored eggs. How many eggs are there altogether?

- (A) 13
- (B) 12
- (C) 15
- (D) 17

19 For a birthday party, Shyam blew up 13 balloons and Steffi blew up 3 balloons. How many balloons were blown up in total by Shyam and Steffi?

- (A) 13
- (B) 12
- (C) 16
- (D) 7

20 There are 8 fishes in a fish tank. After some time, 5 more fishes were added. Again, after some time, 4 more fishes were added. How many fishes are there in total?

- (A) 13
- (B) 12
- (C) 16
- (D) 17

Next Section: Chapter Review

ADDITION WITHIN 20

Chapter Review 1.5

1 Count and write the total number of objects.

_____ butterflies + _____ butterflies = _____ butterflies in total.

2 At the party, 5 guests entered during the first hour and 10 more guests entered during the second hour. How many guests entered in total?

A) 15   B) 13   C) 17   D) 12

3 There are 9 airplanes at an airport. 4 more airplanes land at the same airport. How many airplanes are at the airport?

A) 15   B) 13   C) 17   D) 12

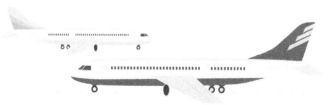

ADDITION WITHIN 20

 1.5   **Chapter Review**

4   Which addition sentence matches the picture?

(A) 4 + 8 = 12   (B) 8 + 4 = 12   (C) 6 + 6 = 12   (D) 7 + 5 = 12

5   Jessy picked 2 oranges from one tree, she picked 5 oranges from the next tree, and 4 oranges from the last tree. How many oranges did she pick in all?

(A) 15   (B) 13   (C) 11   (D) 12

6   There are 8 blue bikes, 3 red bikes, and 3 green bikes locked to the bike rack. How many bikes are on the bike rack in all?

(A) 15   (B) 13   (C) 11   (D) 14

7   Fill in the blank. One possible equation to match this picture is 10 + _____ = _____.

ADDITION WITHIN 20

Chapter Review 1.5

8 Fill in the number sentences and write the 10+ facts.
7 + 8 = _____ , 10 + _____ = _____ .

9 Count on the number line to solve the equation 7 + 5 = _____ .

0 1 2 3 4 5 6 7 8 9 10 11 12 13

10 There are 9 cars, 6 bikes and 4 buses in the parking area. Which equation shows how many vehicles are there in the parking lot.

(A) 9 + 4 + 3 = ?   (B) ? + 6 + 2 = ?
(C) 9 + 6 + 4 = ?   (D) 9 − 6 + 4 = ?

11 Which addition equations could represent the picture?

(A) 6 − 3 = ?   (B) 6 + 3 = ?
(C) 9 − 3 = ?   (D) 9 + 3 = ?

12 Which number goes in the blank? 10 + _____ = 5 + 4 + 9.

(A) 6   (B) 7   (C) 8   (D) 9

ADDITION WITHIN 20

1.5 Chapter Review

13 Count and write the total number of candies.

   +

_____ candies + _____ candies = _____ candies in total.

14 Which number goes in both blanks to make the equations true?
5 + _____ = 12

Ⓐ 7   Ⓑ 8   Ⓒ 6   Ⓓ 9

15 Use the picture to solve the equation: 10 + _____ = 11

ADDITION WITHIN 20

Chapter Review 1.5

16 Use the number line to solve the equation 9 + 7 = _____ .

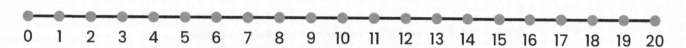

17 Find the missing number using the picture.

18 Use the ten frames to find the answer. 9 + 3 = _____ .

19 Angel currently has 7 teeth. In two years, she will have grown an additional 6 teeth. How many teeth will she have in total after two years?

_____

ADDITION WITHIN 20

1.5 Chapter Review

20. Count and write the total number of objects.

_____ flowers + _____ flowers = _____ flowers in all.

Next Chapter: Subtraction Within 20

CHAPTER 2

SUBTRACTION WITHIN 20

DOT-TO-DOT UNICORN

SUBTRACTION WITHIN 20

SUBTRACTION BASICS

Subtraction is a fundamental mathematical operation used to find the difference between two numbers. It involves taking away or removing a certain quantity from another quantity. Subtraction is denoted by the minus sign (-).

Subtraction is all about finding out "how many more" you need to reach a certain number or "how many left" after taking away some items.

In subtraction, we have three main components:

Minuend: This is the number from which the other number will be subtracted.

Subtrahend: This is the number that is subtracted from the minuend.

Difference: This is the result obtained after subtracting the subtrahend from the minuend.

To perform subtraction, we start with the minuend and subtract the subtrahend from it. The resulting value is the difference.

For example, in the subtraction problem 8 - 5 = 3, 8 is the minuend, 5 is the subtrahend, and 3 is the difference. It means that if we take away 5 from 9, we are left with 4.

Minuend 8 - 5 = 3 Difference
Subtrahend

SUBTRACTION WITHIN 20

2.1 Subtraction Basics

1. Jessica has 18 crayons with her. She gave 5 crayons to her friend. How many crayons does Jessica have now?

A) 13   B) 10   C) 9   D) 12

2. Jessy went to a party. There were 18 cakes at the party. 6 cakes were eaten. How many cakes were left?
She made this drawing with a cake to help her solve the problem: What should she do next?

A) Draw 3 more cakes.
B) Draw 1 more cake.
C) Cross out 16 cakes.
D) Cross out 6 cakes.

3. John wants to save $15.00 to buy a new car toy. He has saved $5.00 so far. How many more dollars does he need to save to have $15.00?

A) $6   B) $8   C) $10   D) $9

4. Jerry has 12 apples with him. He gave 3 apples to his sister. How many apples does he have with him?

A) 8   B) 7   C) 9   D) 10

SUBTRACTION WITHIN 20

Subtraction Basics — 2.1

5 What is  minus ? _____.

6 Which number goes in the blank to make the equation true?

9 - ☐ = 5

Ⓐ 5   Ⓑ 2   Ⓒ 4   Ⓓ 3

7 Robin picked 9 roses from the garden. He gave 3 roses to his mom. How many roses does he still have?

Ⓐ 4   Ⓑ 9   Ⓒ 5   Ⓓ 6

8 Which subtraction equation could represent the picture?

Ⓐ 9+3=?   Ⓑ 3-9=?   Ⓒ 9-3=?   Ⓓ 12-3=?

SUBTRACTION WITHIN 20

2.1 Subtraction Basics

9. Which equation is part of the same fact family as 6 + ____ = 10?

(A) 4 + 10 = ?   (B) 10 − 6 = ?   (C) 6 − 3 = ?   (D) 11 + 2 = ?

10. Use the picture to help you complete the subtraction equation.
10 − 3 = _____.

11. Jerry has 10 fishes in the aquarium. He gave 4 fishes to his friend as a birthday gift. How many fishes remain in the aquarium?

(A) 5   (B) 6   (C) 7   (D) 4

12. Rachel bought 15 lollipops. She ate 4 lollipops. How many lollipops does she have now?

(A) 11   (B) 10
(C) 12   (D) 13

SUBTRACTION WITHIN 20

Subtraction Basics — 2.1

13. Peter caught 8 butterflies in his backyard. He let 3 of them go. How many butterflies does he still have?

- (A) 4
- (B) 5
- (C) 6
- (D) 2

14. Ryan had 12 drawings. He gave some to his brother. Now he has 5. How many drawings did Ryan give his brother? 12 - ____ = 5.

- (A) 8
- (B) 5
- (C) 9
- (D) 7

15. Use the picture to help you solve the equation below. Fill in the blank. 10 - ____ = 7

- (A) 3
- (B) 9
- (C) 11
- (D) 12

16. What is  minus ____?

SUBTRACTION WITHIN 20

2.1 Subtraction Basics

17 Jessy has 8 Barbie toys. She gave 2 barbie toys to her friend. How many toys does she still have?

(A) 4   (B) 9   (C) 5   (D) 6

18 Which subtraction equation is part of the same fact family as 7 + 3 = 10

(A) 4 + 10 = 14   (B) 10 − 7 = 3
(C) 10 + 3 = 13   (D) 11 − 2 = 9

19 Which subtraction equation could represent the picture?

(A) 6 + 3 = ?   (B) 3 − 6 = ?
(C) 6 − 3 = ?   (D) 9 − 3 = ?

SUBTRACTION WITHIN 20

Subtraction Basics 2.1

20 Which subtraction equation could represent the picture?

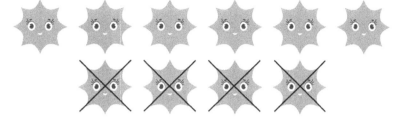

Fill in the blank. 10 - _____ = 6

Ⓐ 4   Ⓑ 5   Ⓒ 6   Ⓓ 2

Next Section: Subtraction by Decomposing a Number into Ten

SUBTRACTION WITHIN 20

SUBTRACTION BY DECOMPOSING A NUMBER INTO TEN

Subtraction by decomposing a number into ten is a mental math strategy that involves changing a number to ten and then adjusting the answer. When you break a number apart, it's called decomposing. Breaking Apart, also known as "Separating" or. "Decomposing" Break both numbers down to place value and add each, starting with the.

Take from ten strategy: A strategy that involves breaking apart the larger number before subtracting from a unit of ten.

Example:

Subtract by taking from ten strategies.

$$11 - 8$$
$$11 - 8 = 10 + 1 - 8$$
$$= 10 - 8 + 1$$
$$= 2 + 1$$
$$= 3$$

SUBTRACTION WITHIN 20

Subtraction by Decomposing a Number into Ten — 2.2

1. Simon bought some marbles. He gave $18 and the shopkeeper gave him $11. What is the cost of the marbles? 18 – 10 – 1 = _____.

- (A) $ 7
- (B) $ 6
- (C) $ 5
- (D) $ 8

2. Emily purchased some flowers. She gave $14 and the shopkeeper gave her $10. What is the cost of the flower?

14 – _____ = _____.

3. There are 15 mangoes in a basket and 7 are rotten. How many mangoes are good? 15 – 5 – 2 = _____.

- (A) 7
- (B) 6
- (C) 8
- (D) 5

4. David bought a shirt for $10. He gave the cashier $19. How much money did he get back?

- (A) $ 7
- (B) $ 6
- (C) $ 8
- (D) $ 9

5. Sam bought a chocolate box for $10. He paid with $11. How much change did he get?

11 – _____ = _____.

- (A) $ 5
- (B) $ 3
- (C) $ 1
- (D) $ 0

SUBTRACTION WITHIN 20

2.2 Subtraction by Decomposing a Number into Ten

6 There are 18 apples in a basket and 9 are rotten. How many apples are good?

_____

7 Tom bought some bananas. He gave $15 and the shopkeeper gave him $12. What is the cost of the bananas?

15 - _____ = _____.

8 There are 16 cherries in a basket and 7 are rotten. How many cherries are good?

16 - _____ = _____.

9 Jerry bought some roses. He gave $11 and the shopkeeper gave him $7. What is the cost of the roses?

11 - _____ = _____.

SUBTRACTION WITHIN 20

Subtraction by Decomposing a Number into Ten — 2.2

10. Peter bought a toy car for $12. He paid with $17. How much change did he get?

17 - _____ = _____.

11. Mary has 12 wooden rods, 4 are broken. How many are not broken?

12 - _____ = _____.

12. Jack has 15 cookies, 6 are broken. How many cookies are not broken?

15 - _____ = _____.

(A) 7   (B) 6
(C) 5   (D) 9

13. There were 15 flowers in a plant. Sam picked 11 of them. How many flowers are left on the plant now? 15 - _____ = _____.

(A) 4   (B) 6
(C) 5   (D) 9

SUBTRACTION WITHIN 20

Subtraction by Decomposing a Number into Ten 2.2

14 Helen has 14 balloons, 5 are broken. How many are not broken?

14 – _____ = _____ .

15 Complete the subtraction.

15 – 9 = _____ .

16 Complete the subtraction.

10 – 7 = _____ .

17 Roy bought 11 toy ducks; He gave 3 toy ducks to her sister? Complete the number sentence by decomposing a number leading to ten.

11 – 3 = _____ .

SUBTRACTION WITHIN 20

Subtraction by Decomposing a Number into Ten — 2.2

18. James scored 5 points less than the score of Tom. If Tom scored 14 points, what did James score?

14 - _____ = _____ .

(A) 7   (B) 6   (C) 5   (D) 9

19. Max had 11 kiwis. He ate 5 kiwis. How many kiwis did he have?

11 - _____ = _____ .

(A) 3   (B) 4   (C) 5   (D) 6

20. Jerry bought 9 oranges. He gave 3 oranges to his neighbors. He wanted to give the remaining oranges to his mother. How many oranges did he give to his mother?

9 - _____ = _____ .

(A) 3   (B) 4   (C) 5   (D) 6

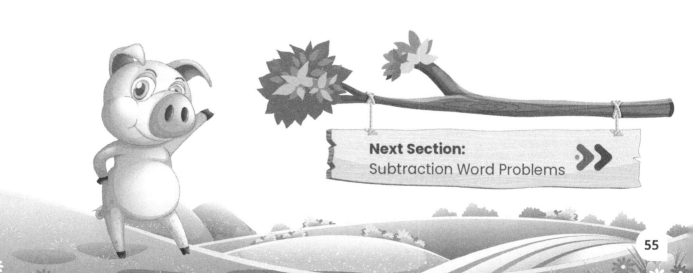

Next Section: Subtraction Word Problems

SUBTRACTION WITHIN 20

SUBTRACTION WORD PROBLEMS

Subtraction is nothing but "take away" or "how many are left" or "how many more or less."

Steps for solving subtraction word problems:

1. Read the problem carefully
2. Identify the minuend and the subtrahend
3. Solve the problem

Examples:

Emily has five games and Marson has seven games. How many fewer games does Emily have?

Fewer games for Emily = 7 − 5 = 2 games

(Here, minuend is 7 and subtrahend is 5)

SUBTRACTION WITHIN 20

Subtraction Word Problems — 2.3

1. Reena has 8 coins. Robin has 2 coins. How many more coins does Reena have?

Ⓐ 6   Ⓑ 7   Ⓒ 5   Ⓓ 4

2. Joey had 10 car toys. He lost 3 of them. How many car toys does Joey have left?

Ⓐ 6   Ⓑ 7   Ⓒ 3   Ⓓ 9

3. Andria bought 17 gift boxes to distribute to his relatives. She distributed 9 gift boxes. How many gifts does she have left?

Ⓐ 4   Ⓑ 3   Ⓒ 8   Ⓓ 5

4. There are 10 beads in the box. 2 beads are green in color, 4 beads are blue in color, and the rest of the beads are red in color. How many are red beads?

Ⓐ 4   Ⓑ 3   Ⓒ 7   Ⓓ 5

SUBTRACTION WITHIN 20

2.3 Subtraction Word Problems

5. Maria had 13 cookies. She gave 3 cookies to his friends. How many cookies does Maria have now?

- A) 8
- B) 9
- C) 7
- D) 10

6. Christina bought 11 star stickers. She kept 3 star stickers, and then gave the rest to her friends. How many start stickers did she give away?

---

7. Angel wants a new toy that costs $18.00. She has saved $6.00. How many more dollars does she need to save?

- A) $6
- B) $8
- C) $12
- D) $9

8. Tom has 6 car toys. His friend came over with some more car toys. Together, they have 10 car toys. How many car toys did his friend bring over?

- A) 4
- B) 3
- C) 2
- D) 5

SUBTRACTION WITHIN 20

Subtraction Word Problems 2.3

9 Mouna had 10 strawberries. She ate 3 in the morning. She ate rest in the evening. How many strawberries did she eat in the evening?

(A) 4 (B) 6 (C) 5 (D) 7

10 Choose the correct answer:

12 people went to the exhibition. There were 8 boys and the rest were girls. How many girls are there?

(A) There are four girls.   (B) There are five girls.

(C) There are six girls.   (D) There are seven girls.

11 Fill in the blanks.

Charles bought 4 apples from the shop. His mom gave him _____ apples. Now he has 7 apples.

12 Jerry went to a party. There were 12 burgers at the party. 5 burgers were eaten. How many burgers were left?
He made this drawing, to help him solve the problem:

What should he do next?

(A) Draw 3 more burgers.   (B) Draw 1 more burger.

(C) Cross out 12 burgers.   (D) Cross out 5 burgers.

SUBTRACTION WITHIN 20

2.3   Subtraction Word Problems

13 Sarah has 4 video games and Sam has 10 video games. How many more video games does Sam have?

  (A) 4   (B) 6   (C) 8   (D) 5

14 Tim bought 5 parrots and 11 lovebirds on Wednesday. How many more lovebirds are there than the parrots?

  (A) 8   (B) 4   (C) 6   (D) 7

15 There are 25 students in a class with 15 boys and 10 girls.. How many more boys are there when compared with girls?

  (A) 6   (B) 7   (C) 8   (D) 5

16 Mary paid $8 for her lunch. Steffi paid $13 for her lunch. How much more did Steffi pay for her lunch?

  (A) $5   (B) $6   (C) $7   (D) $8

17 During the math test, Pinky scored 12 marks, and Tim scored 19 marks. How many marks does Pinky need to score to reach Tim's marks?

  (A) 6   (B) 8   (C) 7   (D) 9

SUBTRACTION WITHIN 20

Subtraction Word Problems 2.3

18

Sam	
Linton	

The figure shows the number of mangoes Sam and Linton have. How many more mangoes does Sam have?

(A) 4 (B) 5 (C) 6 (D) 7

19 Simon had 15 pancakes. He gave some to his brother. Now he has 7. How many pancakes did Simon give his brother?

15 - _____ = 7

(A) 8 (B) 5 (C) 9 (D) 7

20 Angel planted 10 trees and Blessy planted 19 trees. How many more trees were planted by Blessy than Angel?

(A) 7 (B) 5 (C) 4 (D) 9

Next Section: Chapter Review

SUBTRACTION WITHIN 20

2.4 Chapter Review

1. Brito bought 13 erasers. He gave 9 erasers to his friend. How many erasers does he still have?

(A) 4 (B) 5 (C) 6 (D) 7

2. Find the difference:

_____ cupcakes - _____ cupcakes = _____ cupcakes.

3. Mike had 14 red balls and 9 yellow balls. How many red balls did Mike have more than yellow balls?

(A) 8 (B) 5 (C) 9 (D) 7

4. Which number makes the equation true? 15 - _____ = 6

(A) 9 (B) 5 (C) 7 (D) 4

SUBTRACTION WITHIN 20

Chapter Review   2.4

5   Which subtraction equation could represent the picture?

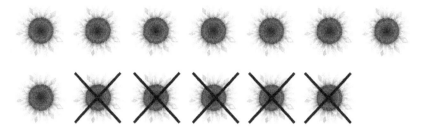

Ⓐ 8 + 5 = ?   Ⓑ 4 − 9 = ?   Ⓒ 9 − 4 = ?   Ⓓ 13 − 5 = ?

6   Pinky and Simon Had 12 tickets with them. Pinky sold 3 tickets and Simon sold 9 tickets. How many fewer tickets are sold by pinky than simon ?

Ⓐ 9   Ⓑ 10   Ⓒ 6   Ⓓ 7

7   Which number goes in the blank? Use the number line.

11 − 6 = _____.

0 1 2 3 4 5 6 7 8 9 10 11 12 13 14 15 16 17 18 19 20

8   Use the picture to help you finish the subtraction equation.

9 − 2 = _____ .

SUBTRACTION WITHIN 20

2.4 Chapter Review

9 Which equation is false?

A) $10 - 4 = 6$ B) $12 - 5 = 7$ C) $4 + 10 = 14$ D) $11 - 4 = 14$

10 Write a related number sentence that matches the number sentence.

$10 - 3 =$ _____.

11 Fluffy caught 10 butterflies. Puffy caught 17 butterflies. How many more butterflies did Puffy catch compared to Fluffy?

A) 8   B) 9   C) 5   D) 7

12 Which number goes in the blank? Use the number line.

0 1 2 3 4 5 6 7 8 9 10 11 12 13 14 15 16 17 18 19 20

$18 - 5 =$ _____.

13 Use the picture to help you solve the equation below.

Fill in the blank.

$10 -$ _____ $= 4$

SUBTRACTION WITHIN 20

Chapter Review 2.4

14. Edwin took 13 steps and Allen took 5 steps in 1 minute. How many steps were taken by Allen than Edwin?

(A) 8 (B) 5 (C) 6 (D) 7

15. John bought 12 balloons. He gave 4 balloons to his neighbors. He wanted to give the remaining balloons to his friend. How many balloons did he give to his friend?

_____ - 4 = _____ = 10 - _____ .

16. Which number goes in the blanks to make the equations true?

18 - _____ = 7

(A) 9 (B) 5 (C) 7 (D) 11

17. Jonny needs to put 19 color pencils on the table. He put 6 color pencils on the table already. How many more color pencils does he need to put on the table?

(A) 15 (B) 10 (C) 13 (D) 11

SUBTRACTION WITHIN 20

2.4 Chapter Review

18 Tim has 5 cotton candies and Laura has 14 cotton candies. How many fewer cotton candies does Tim have than Laura?

14 - _____ = _____.

19 Which number goes in the blank? Use number line.

14 - 6 = _____.

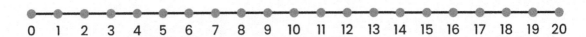

20 Tom bought 9 strawberries. He gave 3 strawberries to his neighbors. He wanted to give the remaining strawberries to his mother. How many did he give to his mother?

9 - _____ = _____.

Ⓐ 3   Ⓑ 4   Ⓒ 5   Ⓓ 6

Next Chapter: Properties of Additive and Subtractive Operations and Equations

CHAPTER 3
PROPERTIES OF ADDITIVE AND SUBTRACTIVE OPERATIONS AND EQUATIONS

DOT-TO-DOT BABY GIRAFFE

PROPERTIES OF ADDITIVE AND SUBTRACTIVE OPERATIONS AND EQUATIONS

COMMUTATIVE PROPERTY AND ASSOCIATIVE PROPERTY OF ADDITION

COMMUTATIVE PROPERTY

The commutative property is a fundamental concept in mathematics, specifically in the field of algebra. It refers to the property of certain mathematical operations that allows the order of the operands to be changed without affecting the result.

The commutative property of addition states that the order in which numbers are added does not affect the sum.

You can add two numbers in any order and still obtain the same result.

For example, 2 + 3 = 3 + 2 = 5.

The commutative property holds true for any number of addends. For instance, 2 + 3 + 4 = 4 + 3 + 2 = 9.

COMMUTATIVE PROPERTY AND ASSOCIATIVE PROPERTY OF ADDITION

ASSOCIATIVE PROPERTY

The associative property of addition is a fundamental property in mathematics that states that the grouping of numbers being added does not affect the sum.

The associative property of addition states that the grouping of numbers being added does not affect the sum.

Adding three or more numbers, you can regroup them without changing the sum.

For example, $(2 + 3) + 4 = 2 + (3 + 4) = 9$.

The associative property holds true for any number of addends. For instance, $(2 + 3) + 4 + 5 = 2 + (3 + 4 + 5) = 14$.

PROPERTIES OF ADDITIVE AND SUBTRACTIVE OPERATIONS AND EQUATIONS

Commutative Property and Associative Property of Addition — 3.1

1 Write the numbered sentences to match the objects.

_____ mangoes + _____ strawberries = _____ pieces of fruit
= _____ strawberries + _____ mangoes.

2 Peter bought 5 white shirts and 8 black shirts for an event. How many shirts did he buy altogether? 5 + 8 = _____ .

A) 14   B) 13   C) 15   D) 12

3 Write the number sentences to match the objects.

_____ butterflies + _____ bees = _____ insects
= _____ bees + _____ butterflies.

4 Jill wants to pick an equation that shows the Commutative Property of Addition for 9 + 7 = 16. Which equation should she pick?

A) 8 + 5 = 13   B) 7 + 2 = 9   C) 4 + 9 = 13   D) 7 + 9 = 16

71

PROPERTIES OF ADDITIVE AND SUBTRACTIVE OPERATIONS AND EQUATIONS

 3.1 Commutative Property and Associative Property of Addition

 Quick Tips

Hint: Start from the larger number.

Example: 2 + 8

▶ 2 + 8 is Harder: "2 ... 3, 4, 5, 6, 7, 8, 9, 10"

▶ 8 + 2 is Easier: "8 ... 9, 10"

▶ So do 8 + 2 instead

You will get the same answer because of the Commutative Property of Addition.

5 Jessy has 8 gold fishes and 6 silver fishes in the fish tank. How many fishes does she have in all?

8 + 6 = _____ .

6 Claire has 9 eggs in the refrigerator and 6 eggs outside. How many eggs are there in all? 9 + 6 = _____ .

Ⓐ 14   Ⓑ 13   Ⓒ 15   Ⓓ 12

7 Tim bought 8 sunflowers and 6 roses. How many flowers does Tim buy in all? 8 + 6 = _____ .

Ⓐ 14   Ⓑ 13   Ⓒ 15   Ⓓ 12

PROPERTIES OF ADDITIVE AND SUBTRACTIVE OPERATIONS AND EQUATIONS

Commutative Property and Associative Property of Addition — 3.1

8 Which choice would show the commutative property of addition for 3 + 6 = _____?

- A) 6 + 3 = _____
- B) 5 + 3 = _____
- C) _____ = 4 + 9
- D) _____ = 7 + 9

9 Use the picture and write all possible answers.

_____

10 Write 2 number sentence to match the number bond.

7   5

_____ + _____ = _____ .

11 Tom bought 7 red roses and 11 pink roses for the function. How many roses did he buy altogether? _____ + _____ = _____ .

- A) 18
- B) 13
- C) 15
- D) 12

PROPERTIES OF ADDITIVE AND SUBTRACTIVE OPERATIONS AND EQUATIONS

3.1 Commutative Property and Associative Property of Addition

12 At a party, Rita receives 3 dolls, 2 teddy bears, and 2 toy cars. How many toys did she receive in all? 3 + 2 + 2 = _____.

Ⓐ 7 Ⓑ 6 Ⓒ 9 Ⓓ 8

13 Blessy has, 7 sparrows, 5 parrots, and 7 lovebirds. How many pets does she have?

☐ + ☐ + ☐ = ☐

14 Choose the best answer.
Ben worked 8 hours per day. Calro worked 5 hours per day, and Nick worked 3 hours per day. How many hours did they work altogether?
_____ + _____ + _____ = _____ .

Ⓐ 14 Ⓑ 16 Ⓒ 15 Ⓓ 12

15 John has 5 red apples and 7 green apples in the basket. How many apples does he have in all? _____ + _____ = _____ .

☐ + ☐ = ☐   ☐ + ☐ = ☐

PROPERTIES OF ADDITIVE AND SUBTRACTIVE OPERATIONS AND EQUATIONS

Commutative Property and Associative Property of Addition — 3.1

16 Maria has, 8 blue pens, 5 black pens, and 3 red pens. How many pens does she have? Write 2 addition sentences and solve.

☐ + ☐ + ☐ = ☐

☐ = ☐ + ☐ + ☐

17 Write two number sentences and solve.

Roy bought 6 apples, 3 oranges, and 1 watermelon. How many fruits did he buy?

☐ + ☐ + ☐ = ☐

☐ = ☐ + ☐ + ☐

18 Choose the best answer.

In a garden, there are 7 red vinca plants, 3 pink vinca plants, and 8 white vinca plants. How many plants are in the garden?

7 + 3 + 8 = _____.

Ⓐ 14   Ⓑ 13   Ⓒ 18   Ⓓ 12

PROPERTIES OF ADDITIVE AND SUBTRACTIVE OPERATIONS AND EQUATIONS

3.1 Commutative Property and Associative Property of Addition

19 Emy purchased 9 tomatoes and 8 cabbages. How many vegetables did she buy altogether? Write two addition sentences and solve.

_____ + _____ = _____ .

_____ + _____ = _____ .

20 Count and complete the sentence.

____ balloons + ____ Heart balloons + ____ Star balloons = ____ balloons

What is another way to solve this problem? Write another addition sentence and solve.

Next Section: Three Number Addition or Subtraction

PROPERTIES OF ADDITIVE AND SUBTRACTIVE OPERATIONS AND EQUATIONS

THREE NUMBER ADDITION OR SUBTRACTION

ADDITION

Adding three numbers means combining three numerical values to find a total.

Examples:

$$\text{Add } 2 + 3 + 1$$
$$2 + 3 + 1 = 5 + 1$$
$$= 6$$

SUBTRACTION

Subtracting three numbers involves taking away one numerical value from the sum of the other two numerical values to find the difference.

Examples:

$$\text{Subtract } 7 - 2 - 3$$
$$7 - 2 - 3 = 5 - 3$$
$$= 2$$

PROPERTIES OF ADDITIVE AND SUBTRACTIVE OPERATIONS AND EQUATIONS

3.2 Three Number Addition or Subtraction

1 Count and complete the number sentence.

_____ Broccoli + _____ Capsicums + _____ Beetroots = _____ Vegetables.

2 Maria cooked 4 eggs in the morning and 8 eggs in the afternoon. She cooked 3 eggs at night. How many eggs did she cook in all?

A) 15   B) 8   C) 10   D) 11

3 James planted 9 rose plants and 6 jasmine plants in the front garden. Then he moved two rose plants to the back garden. How many plants are in the front garden?

☐ + ☐ - ☐ = ☐

4 Tina baked 5 cookies in the morning and 9 cookies in the evening. She ate 3 cookies at night. How many cookies did she have now?

A) 19   B) 11   C) 10   D) 17

PROPERTIES OF ADDITIVE AND SUBTRACTIVE OPERATIONS AND EQUATIONS

Three Number Addition or Subtraction — 3.2

5 Jack has 8 red fishes and 7 black fishes in a fish tank. He removed 1 fish from a tank. How many fishes are in a tank?

☐ + ☐ − ☐ = ☐

6 Count and complete the equation.

_____ Cabbages − _____ tomatoes + _____ carrots = _____ Vegetables.

7 There are 5 pink lotus flowers, 2 white lotus flowers, and 4 white lilies in the pond. How many flowers are there in all?

A) 11   B) 7   C) 12   D) 13

8 Ben has 8 blue pens and 5 black pens. He gave away 3 pens. How many pens does he have now?

A) 9   B) 8   C) 10   D) 11

PROPERTIES OF ADDITIVE AND SUBTRACTIVE OPERATIONS AND EQUATIONS

3.2 Three Number Addition or Subtraction

9. Complete the number sentence.
8 + 3 − 4 = _____ , 10 − 3 = _____ .

- (A) 6
- (B) 7
- (C) 5
- (D) 8

10. Melissa has 5 candies, 8 cookies, and 5 chocolates. How many items does she have in all?

- (A) 19
- (B) 20
- (C) 18
- (D) 17

11. Complete the number sentence.
7 + 4 − 2 = _____ , 10 − 1 = _____ .

- (A) 9
- (B) 8
- (C) 10
- (D) 7

12. Solve. 3 + 2 + 5 = _____ .

- (A) 5
- (B) 7
- (C) 10
- (D) 8

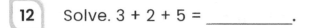

PROPERTIES OF ADDITIVE AND SUBTRACTIVE OPERATIONS AND EQUATIONS

Three Number Addition or Subtraction — 3.2

13 Count and complete the sentences.

_____ apples + _____ oranges + _____ strawberries = _____ pieces of fruit.

14 In a tree, there are 9 parrots and 3 pigeons. Four of them flew away. How many birds are still on the tree?

A) 11   B) 8   C) 7   D) 13

15 Jeo saw 8 lions, 4 tigers, and 5 deer at the zoo. How many animals did he see?

A) 15   B) 11   C) 13   D) 17

16 Alan has 7 green color pencils and 9 blue color pencils. He misplaced 4 pencils. How many pencils does he have now?

A) 12   B) 10   C) 17   D) 13

PROPERTIES OF ADDITIVE AND SUBTRACTIVE OPERATIONS AND EQUATIONS

3.2 Three Number Addition or Subtraction

17 Which choice is equal to 5 + 2 + 5 = ?

A) 10+3=___   B) ___ = 10+6   C) 10+0=___   D) ___ = 10+2

18 Ronald bought marbles. He bought 7 pink marbles, 5 green marbles, and 5 yellow marbles. How many marbles did he buy in all?

☐ + ☐ + ☐ = ☐

19 Joseph made 8 red kites and 4 yellow kites. He gave 3 to his friend. How many kites does he have now?

A) 11   B) 9   C) 7   D) 8

20 Mercy has 6 shirts, 3 pants and 5 gowns. How many items did she have in all?

☐ + ☐ + ☐ = ☐

Next Chapter: Add or Subtract Within 100 Using Properties of Operations

PROPERTIES OF ADDITIVE AND SUBTRACTIVE OPERATIONS AND EQUATIONS

ADD OR SUBTRACT WITHIN 100 USING PROPERTIES OF OPERATIONS

Properties of operations that relate addition and subtraction, commutative property, and associative properties.

Commutative property of addition:
When two numbers are added, the sum is the same regardless of the order in which the numbers are added.

Example:
$$40 + 50 = 50 + 40 = 90$$

Associative property of addition:
When three or more numbers are added, the sum is the same regardless of the way in which the numbers are grouped.

Example:
$$20 + (40 + 10) = (20 + 40) + 10 = 70$$

PROPERTIES OF ADDITIVE AND SUBTRACTIVE OPERATIONS AND EQUATIONS

3.3 Add or Subtract Within 100 Using Properties of Operations

1 There are 52 children standing in a line. 25 children joined them. How many children are in the line?

(A) 77   (B) 82   (C) 72   (D) 89

2 There are 48 mangoes in the basket. Allan added 21 mangoes in the same basket. How many mangoes are in the basket?

_____

3 Stella has 10 cupcakes with her. She bought eight more cupcakes. How many cupcakes does she have in all?

(A) 12   (B) 18   (C) 28   (D) 22

4 Choose the correct answer and fill in the blank.

$$70 - 56 = \underline{\qquad}.$$

(A) 12   (B) 14   (C) 24   (D) 18

5 Mark has 2 sticks of ten cubes. His mom gave him 9 ones. How many cubes does he have in all?

(A) 23   (B) 25   (C) 27   (D) 29

PROPERTIES OF ADDITIVE AND SUBTRACTIVE OPERATIONS AND EQUATIONS

Add or Subtract Within 100 Using Properties of Operations — 3.3

6 How many bundles of 10 can Jenny make it with her stars?

Jenny can make _____ tens.

7 Fill in the blanks with 10 less and 10 more.

_____ , 45 , 62 , _____ .

8 April is counting her stamp collection. She put them in piles of 10. She has 4 piles of 10 and 3 loose stamps left over. She gave 1 pile and 3 loose stamps to her friend.
How many stamps are in April now?

A) 30 B) 25 C) 20 D) 28

9 Fill in the blanks with 10 less and 10 more.

_____ , 24 , 34 , _____ .

PROPERTIES OF ADDITIVE AND SUBTRACTIVE OPERATIONS AND EQUATIONS

 3.3 Add or Subtract Within 100 Using Properties of Operations

10 True or False.
88 is same as 9 tens 2 ones.

Ⓐ True Ⓑ False

11 What number is two tens more than 4 tens and 6 ones?

_____

12 Sharon cultivated some vegetables in her garden. There are 38 carrots and 26 cabbages. How many vegetables did she cultivate?

Ⓐ 64   Ⓑ 69   Ⓒ 72   Ⓓ 78

13 Sam wrote a short story book. He writes 17 pages on Monday and 20 pages on Tuesday. How many pages does he write in all?

_____

PROPERTIES OF ADDITIVE AND SUBTRACTIVE OPERATIONS AND EQUATIONS

Add or Subtract Within 100 Using Properties of Operations — 3.3

14 Write the number as tens and ones.
6 tens and 10 ones − 3 tens and 6 ones = _____ tens and _____ ones.

15 Choose the correct answer and fill in the blank.
53 − 31 = _____ .

Ⓐ 12 Ⓑ 24 Ⓒ 22 Ⓓ 18

16 At the beginning of the year Ms.Edwin, the gym teacher, had 45 balls. She said that 10 have been lost or popped. How many balls are there now?
There are _____ balls.

17 Write the number as tens and ones.
4 tens and 8 ones − 2 tens and 4 ones = _____ tens and _____ ones.

PROPERTIES OF ADDITIVE AND SUBTRACTIVE OPERATIONS AND EQUATIONS

3.3 Add or Subtract Within 100 Using Properties of Operations

18 What number is two tens more than 5 tens and 1 ones?

- A) 71
- B) 51
- C) 41
- D) 61

19 True or False.
42 is the same as 4 tens 12 ones.

- A) True
- B) False

20 Write the number as tens and ones.
3 tens and 9 ones − 1 tens and 8 ones = _____ tens and _____ ones.

Next Section: Finding Two Values Equality

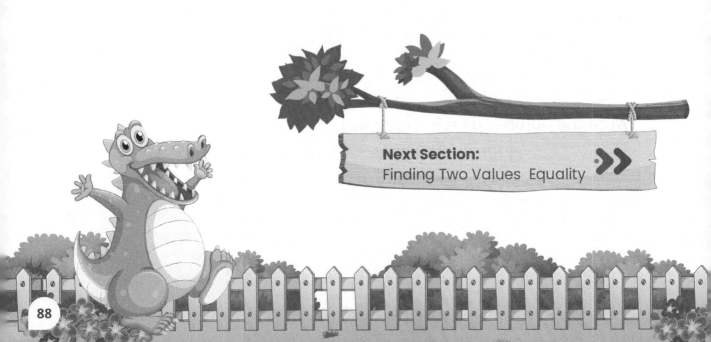

PROPERTIES OF ADDITIVE AND SUBTRACTIVE OPERATIONS AND EQUATIONS

FINDING TWO VALUES ARE EQUAL

The equal sign or equality sign (=) is a mathematical symbol used to indicate equality. The symbol = shows that what is on the left of the sign is equal in value or amount to what is on the right of the sign.

When the Left-Hand Side of the equals sign is the same in value as the Right-Hand Side, then the equation is true.

When the Left-Hand Side of the equals sign is the difference in value as the Right-Hand Side, then the equation is not true or false.

Examples:

3 + 4 = 7 means that 3 + 4 is equal to 7

5 + 8 ≠ 14 means that 5 + 8 is not equal to 14.

PROPERTIES OF ADDITIVE AND SUBTRACTIVE OPERATIONS AND EQUATIONS

3.4 Finding Two Values Are Equal

1 Write the expression that matches the groups of fruit on each box. If they have the same amount of fruit. Write the equal sign between the expressions.

_____ + _____ ☐ _____ + _____

2 Which equation is equal to 7 + 2 = ?

A) 5 + 4 = _____ . B) 3 + 4 = _____ .

C) 8 + 3 = _____ . D) 2 + 9 = _____ .

3 David says that the expressions 15 − 2 and 19 − 6 are not equal. Mike says they are equal. Who is right?

A) David B) Mike

4 What number makes the equation true. 9 + 2 = 3 + _____ .

A) 8 B) 7 C) 9 D) 5

PROPERTIES OF ADDITIVE AND SUBTRACTIVE OPERATIONS AND EQUATIONS

Finding Two Values Are Equal 3.4

5 Solve the equations. Then decide if the equation is true or false.

$$10 + 3 = 15 - 2.$$

(A) True   (B) False

6 What number would make this equation true? $12 - 5 = $ _____.

(A) 8   (B) 6   (C) 7   (D) 9

7 Write the expression that matches the groups of flowers in each box. If they have the same amount of flowers. Write an equal sign between the expressions.

_____ + _____ ☐ _____ + _____

8 Which math equation is equal to $9 + 3 = ?$

(A) $6 + 4 = $ _____.   (B) $6 + 6 = $ _____.

(C) $4 + 3 = $ _____.   (D) $2 + 8 = $ _____.

PROPERTIES OF ADDITIVE AND SUBTRACTIVE OPERATIONS AND EQUATIONS

3.4 Finding Two Values Are Equal

9 Mellisa says that the expressions 11 - 2 and 18 - 9 are not equal. Jessy says they are equal. Who is right?

_____

10 Solve the equations. Then decide if the equation is true or false.
6 + 3 = 19 - 8.

A) True   B) False

11 Solve the equations. Then say if the equation is true or false.
2 + 5 = 3 + 4

A) True   B) False

12 Which answer choice makes the number sentence true?
_____ + 2 = 3 + 4 .

A) 8   B) 7   C) 9   D) 5

PROPERTIES OF ADDITIVE AND SUBTRACTIVE OPERATIONS AND EQUATIONS

Finding Two Values Are Equal 3.4

13 Write the expression that matches the groups of insects in each box. If they have the same number of insects, write an equal sign between the expressions.

_____ + _____ ☐ _____ + _____

14 Which answer choice would make the equation true?
3 + 2 = _____.

(A) 5 + 4 = _____. (B) 3 + 1 = _____.

(C) 4 + 1 = _____. (D) 2 + 4 = _____.

15 Elsie says that the expressions 17 − 5 and 18 − 6 are not equal. Sandra says they are equal. Who is right?

_____

16 Solve the equations. Then decide if the equation is true or false.
3 + 3 = 11 − 5

(A) True (B) False

93

PROPERTIES OF ADDITIVE AND SUBTRACTIVE OPERATIONS AND EQUATIONS

3.4 Finding Two Values Are Equal

17 Which answer choice makes the equation true?

2 + 4 = _____ + 3

- (A) 8
- (B) 3
- (C) 9
- (D) 5

18 Find the missing part to make the number sentences true.

_____ + 7 = 4 + 8

- (A) 18
- (B) 16
- (C) 12
- (D) 17

19 Which subtraction fact is equal to 9 + 2?

- (A) 13 – 2
- (B) 9 – 2
- (C) 11 – 3
- (D) 15 – 2

20 Solve the equations. Then say if the equation is true or false.

4 + 3 = 12 – 6

- (A) True
- (B) False

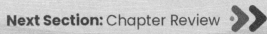

Next Section: Chapter Review

PROPERTIES OF ADDITIVE AND SUBTRACTIVE OPERATIONS AND EQUATIONS

Chapter Review 3.5

1 Write the number sentences to match the objects.

_____ red peppers + _____ onions = _____ vegetables
= _____ onions + _____ red peppers.

2 Which choice equals 6 + 5 = _____?

Ⓐ 7 + 4   Ⓑ 7 + 3   Ⓒ 4 + 9   Ⓓ 7 + 9

3 There are 7 small crabs, 4 big crabs, and 7 black crabs on the beach. How many crabs are on the beach?

☐ + ☐ + ☐ = ☐

4 James bought 7 apples and 5 mangoes from the fruit shop. How many pieces of fruit did James buy in all?

Ⓐ 14   Ⓑ 12   Ⓒ 15   Ⓓ 13

PROPERTIES OF ADDITIVE AND SUBTRACTIVE OPERATIONS AND EQUATIONS

3.5 Chapter Review

5 Write the number sentences to match the number bond.

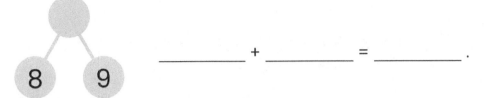

_____ + _____ = _____ .

6 Which addition fact equals 5 + 3 + 9 = _____ ?

A) _____ = 10 + 3   B) 10 + 2 = _____

C) 10 + 5 = _____   D) _____ = 10 + 7

7 Christina bought 5 red balloons, 3 green balloons, and 4 yellow balloons for the party. How many balloons did she buy in all?

A) 12   B) 10
C) 14   D) 11

8 Which addition fact is equal to 8 + 3 = _____ ?

A) 5 + 4 = _____   B) 6 + 5 = _____
C) 3 + 3 = _____   D) 2 + 7 = _____

PROPERTIES OF ADDITIVE AND SUBTRACTIVE OPERATIONS AND EQUATIONS

Chapter Review — 3.5

9 Jenny says that the expressions 11 − 2 and 15 − 6 are not equal. Angel says they are equal. Who is right?

_____

10 Find another way to fix each number sentence to make it true
12 + 3 = _____.

Ⓐ 1+8 = ___   Ⓑ 7+9 = ___   Ⓒ 8+7 = ___   Ⓓ 4+5 = ___

11 Which choice makes the equation true? 12 + 3 = 8 + _____.

Ⓐ 8   Ⓑ 9   Ⓒ 7   Ⓓ 5

12 Write the expression that matches the groups. If they have the same amount of fruit, write the equal sign between the expressions.

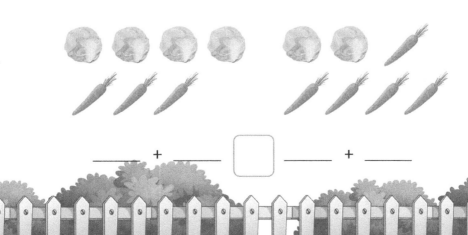

____ + ____ ☐ ____ + ____

PROPERTIES OF ADDITIVE AND SUBTRACTIVE OPERATIONS AND EQUATIONS

3.5 Chapter Review

13 True or false? 12 + 5 = 19 − 2

  (A) True
  (B) False

14 In a garden, there are 7 butterflies and 5 bees. Two of the bees flew away. How many insects are left in the garden now?

  (A) 9
  (B) 8
  (C) 10
  (D) 11

15 Which addition fact is equal to 4 + 7 + 8?

  (A) 10 + 3
  (B) 10 + 9
  (C) 10 + 6
  (D) 10 + 4

16 Maria gets some new dresses for her birthday. She gets 6 pink dresses and 8 blue dresses. She gave 3 of the dresses to her sister. How many dresses does she have now?

☐ + ☐ − ☐ = ☐

PROPERTIES OF ADDITIVE AND SUBTRACTIVE OPERATIONS AND EQUATIONS

Chapter Review 3.5

17 Count and complete the number sentence.

_____ butterflies − _____ bees + _____ bugs = _____ insects.

18 Fill in the blanks.
If 8 + 2 + 3 = 13, then 10 + _____ = 13

Ⓐ 3   Ⓑ 5   Ⓒ 6   Ⓓ 7

19 Which subtraction fact belongs in the same fact family as
11 + _____ = 15?

Ⓐ 15 − 13 = ?   Ⓑ 15 − 11 = ?   Ⓒ 15 + 13 = ?   Ⓓ 13 + 15 = ?

PROPERTIES OF ADDITIVE AND SUBTRACTIVE OPERATIONS AND EQUATIONS

3.5 Chapter Review

20 To change the equation "6 + 5 + 4 = ?" into the equation "10 + 5 = ?", What do you need to do first.

(A) Add 3 and 8
(B) Add 9 and 6
(C) Add 5 and 6
(D) Add 6 and 4

Next Chapter: Place Value

CHAPTER 4
PLACE VALUE

DOT-TO-DOT BABY ELEPHANT

PLACE VALUE

COUNT AND COMPARE NUMBERS WITHIN 120

The most basic strategy to be considered for solving mathematical addition and subtraction problems is to count from one.

Compare the numbers

Compare means to examine the differences between numbers, quantities or values to decide if it is greater than, smaller than, or equal to another quantity. It is good to know if one number is the same as, or smaller than, or bigger than, another number: We use These Signs to compare numbers: = When two values are equal, we use the "equals" sign.

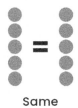
Same

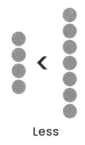
Less

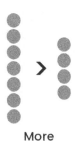
More

Examples:

1. Which is greater? 7 or 9?
Answer: 9 is greater than 7

2. Which is smaller? 8 or 3?
Answer: 3 is smaller than 8

PLACE VALUE

 4.1 Count and Compare Numbers Within 120

1 Count the number of objects and write the number.

_____.

2 Fill in the blanks with the correct numbers.

118 , _____ , _____ , 121.

3 Write the following numbers from least to greatest. 78, 82 , 55

_____

4 Count the number of objects and write the number.

_____.

PLACE VALUE

Count and Compare Numbers Within 120
4.1

5 What number comes before one hundred twenty?

- A) 116
- B) 120
- C) 119
- D) 118

6 Fill in the blanks with the correct number.

115 , _____ , 117 , 118 , _____ .

7 Write the numbers in order from least to greatest. 29, 49 , 17

_____

8 How many more stars are needed to get to 50 stars?

- A) 18
- B) 13
- C) 17
- D) 15

PLACE VALUE

4.1 Count and Compare Numbers Within 120

9 Choose the correct answer.
Blessy decided to count up to one hundred and twenty. She counted to one hundred and fifteen. What number should she say next?

A) 116   B) 120   C) 119   D) 118

10 Fill in the blanks. _____ , 56 , 57 , _____ , 59.

11 Count and compare the balls. Then complete the number sentence using a <, >, or = sign.

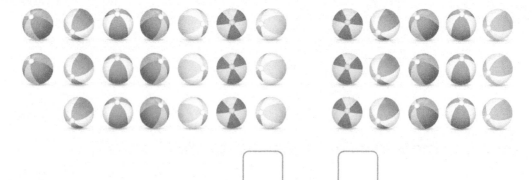

☐ ___ ☐

12 Choose the correct sign to make the number sentence true.

95 _____ 100.

A) <   B) =   C) >

PLACE VALUE

Count and Compare Numbers Within 120   4.1

13. Choose the correct sign to make the number sentence true.

52 _____ 63 _____ 41

14. Count and compare the objects and then fill with <, >, or = sign.

15. Choose the correct symbol to make the number sentence true.

107 _____ 107.

A) <   B) =   C) >

16. Choose the correct answer:

Ronald read 75 pages, and David read 89 pages. Who has read more pages?

A) Ronald   B) David

PLACE VALUE

 4.1 Count and Compare Numbers Within 120

17 Count and compare the objects, and then choose the correct sign: <, >, or =.

18 Which symbol makes the number sentence true? 115 _____ 103.

    Ⓐ <   Ⓑ =   Ⓒ >

19 What number is greater than 45, but less than 60?

    Ⓐ 55   Ⓑ 63   Ⓒ 82   Ⓓ 71

20 Fill in the blanks with <, =, and >.

15 _____ 10.

    Ⓐ <   Ⓑ =   Ⓒ >

Next Section: Tens and Ones

PLACE VALUE

TENS AND ONES

Tens and ones are organizing numbers based on their place value. The tens place represents the second digit from the right, while the ones place represents the rightmost digit.

Examples:

Consider the number 24.

In the two-digit number 24 the first number from the right to the left is 4 which represent 4 single units or equivalently $1 + 1 + 1 + 1 = 4$ ones.

Similarly, in the two-digit number 24, the second number from the right to the left is 2 which represent 2 tens or equivalently $10 + 10 = 20$ or 2 tens.

PLACE VALUE

4.2 Tens and Ones

1 Determine the place value of the underlined digit. <u>7</u>7
_____.

2 How many tens and ones are in the number forty-two?

☐ tens and ☐ ones = 42

3 What place is the underlined digit in? Tens or ones?
4<u>3</u> _____.

4 Ben has two tens. His teacher gives him eight ones. What number does he have?

Ⓐ 28   Ⓑ 23   Ⓒ 15   Ⓓ 18

5 What is 56 minus 2 tens? ____ tens and ____ ones = ____ .

Ⓐ 28   Ⓑ 23   Ⓒ 36   Ⓓ 18

110

PLACE VALUE

Tens and Ones 4.2

6 Combine the tens and ones and form the number that represents the below image.

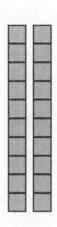

_____

7 What number is in the tens place?

2̲6 - _____

8 How many tens and ones are in the number seventy-nine?

☐ tens and ☐ ones = 79

9 How many tens and ones are in the number fifty-six?

☐ tens and ☐ ones = 56

10 Sarah is counting her cubes. How many bundles of 10 can she make? How do you know?

_____

PLACE VALUE

4.2 Tens and Ones

11 Fill in the blanks. ☐ = 8 tens and 0 ones

12 John has 80 cubes. He gives 5 tens to his partner. How many cubes will he have?

(A) 30 (B) 90 (C) 20 (D) 70

13 How many blocks do I have?

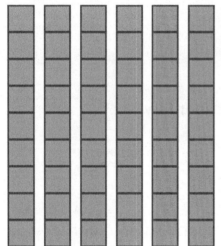

(A) 68 (B) 58
(C) 67 (D) 57

PLACE VALUE

Tens and Ones 4.2

14 Edwin is writing numbers by ones. What numeral should he write after one hundred nine?

He should write _____.

15 How many tens and ones are shown?

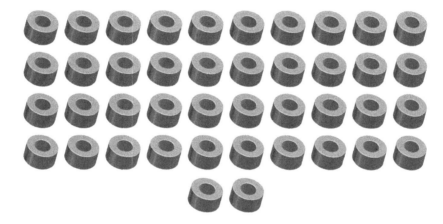

- (A) 4 tens 2 ones
- (B) 5 tens 2 ones
- (C) 4 tens 6 ones
- (D) 5 tens 0 ones

16 Jessy has 70 cubes. She gives 3 tens to her partner. How many cubes will she has?

- (A) 60
- (B) 100
- (C) 40
- (D) 73

PLACE VALUE

4.2 Tens and Ones

17 Choose the correct answer.
The sum of 6 tens and 3 ones, 2 tens and 5 ones is _____ .

(A) 88   (B) 90   (C) 50   (D) 73

18 Rita is thinking of a number that is 20 less than 6 tens and 9 ones. What number is she thinking of? How do you know?

_____

19 Fill in the blanks.

☐ = 3 tens and 1 ones

20 Allen has five tens. His friend gave him nine ones. What number does he have?

(A) 48   (B) 49   (C) 53   (D) 59

Next Section:
Finding Sums and Differences

PLACE VALUE

FINDING SUMS AND DIFFERENCES

Finding sums:

Find the sum of two numbers by adding the numbers together.

Example:

Add a 2-digit number and a 1-digit number

Consider, 42 + 6.

First, write the equation in column form.

$$\begin{array}{r} 42 \\ + 6 \\ \hline \end{array}$$

Add the numbers in the one's place.

$$\begin{array}{r} 42 \\ + 6 \\ \hline 8 \end{array}$$

Then, add the numbers in the tens place.

$$\begin{array}{r} 42 \\ + 6 \\ \hline 48 \end{array}$$

PLACE VALUE

FINDING SUMS AND DIFFERENCES

Finding difference:

Find the difference between the two numbers by subtracting "the numbers in the ones place from the numbers in the ones place" and "the numbers in the tens place from the numbers in the tens place".

The following are the steps to be followed to subtract two-digit numbers.

Step 1: Arrange the numbers vertically one below one, so that the tens' place digits and ones' place digits are lined up. Draw a line under the bottom number to write the respective differences.

Step 2: Subtract the digits in the one's place and write the difference under the line drawn respective to the one's column.

Step 3: Subtract the digits in the tens place and write the difference under the line.

Step 4: The required difference is obtained.

Example:

Subtract 44 ones from 5 tens using the place value chart.

Tens	Ones
5	0
4	3
0	7

Thus, the required difference is 7.

PLACE VALUE

Finding Sums and Differences   4.3

1. James had 76 sunflower stickers with him. He bought 8 more stickers. How many stickers does he have now?

   James has _____ stickers.

2. Sam had 63 balloons with him. He bought 7 more balloons. How many balloons does he have now?

   Sam has _____ balloons.

3. James has rabbits on his farm. He gives the rabbits 50 carrots a day as food. They ate 30 carrots this morning. How many carrots do they have left to eat?

   The rabbits have _____ carrots left to eat.

4. Roy has monkeys on his farm. He gives 60 bananas as food to the monkeys a day. They ate 50 bananas this morning. How many bananas do they have left to eat?

   A) 10   B) 60   C) 80   D) 40

PLACE VALUE

4.3 Finding Sums and Differences

5. Miller had 45 marbles. His sister just gave him 5 more. How many does he have now?

  Ⓐ 30   Ⓑ 50   Ⓒ 47   Ⓓ 49

6. Lucy has 70 flowers. She put 3 tens back. How many flowers does she have now?

_____

7. Ben has two quarters, which he knows equal 50 cents. Then he found 6 cents in his pocket. How much money does he have now?

  Ⓐ 18 cents   Ⓑ 30 cents   Ⓒ 89 cents   Ⓓ 56 cents

8. Mike had 38 base ten blocks. He add 9 more blocks. How many does he have now?

  Ⓐ 50   Ⓑ 52   Ⓒ 47   Ⓓ 49

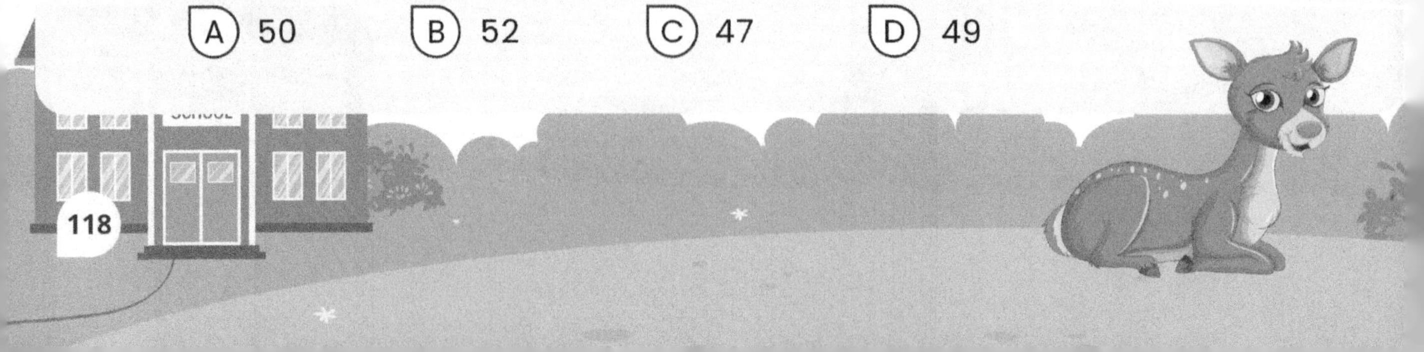

PLACE VALUE

Finding Sums and Differences   4.3

9   Alex had 4 boxes of oranges in his shop. He again bought 8 boxes of oranges to his shop. How many boxes of oranges does Alex have now?

Alex has _____ boxes of Oranges.

(A) 12   (B) 13   (C) 20   (D) 19

10   Jerry has 60 marbles. He put 4 tens back. How many beads does he have now?

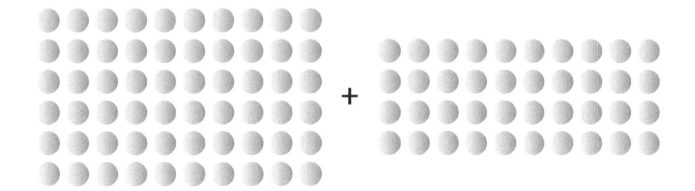

Jerry has _____ marbles.

11   Tim has three quarters, which he knows equal 75 cents. Then he found 4 cents in his pocket. How much money does he have now?

(A) 68 cents   (B) 77 cents   (C) 79 cents   (D) 49 cents

PLACE VALUE

4.3 Finding Sums and Differences

12 Allan has sixty-eight flowers with him. How many bundles of 10 can he make?

Allan can make _____ bundles of 10.

13 How many tens and ones are in the number forty-three? Explain how you know.

_____

14 Ronald has ninety-one apples with him. How many bundles of 10 can he make?

Ronald can make _____ bundles of 10.

15 How many tens and ones are in the number thirty-five? Explain how you know.

_____

PLACE VALUE

Finding Sums and Differences 4.3

16 Peter had 3 boxes of candies. He bought 6 boxes of candies from his friend. How many boxes of candies does Peter have now?

Peter has _____ boxes of candies.

17 Donald has seventy-four fruits with him. How many bundles of 10 can he make?

Donald can make _____ bundles of 10.

18 During recess, 70 First graders are playing on the playground. Ms. Thomson takes her class inside. Ms. Thomson has 20 students. How many students are still on the playground?

A) 50   B) 53   C) 48   D) 61

19 How many tens and ones are in the number twenty-eight? Explain how you know.

_____

PLACE VALUE

4.3 Finding Sums and Differences

20 Ben has fifty-three balls with him. How many bundles of 10 can he make?

Ben can make _____ bundles of 10.

Next Section: Chapter Review

PLACE VALUE

Chapter Review 4.4

1 How many marbles does Silvia have?

Silvia has _____ stickers.

2 Mike is writing his number from 100 to 120. Which number should he write after one hundred and two, when counting by ones?

- A) 103
- B) 105
- C) 107
- D) 108

3 The sum of 7 tens and 6 ones, 2 tens and 2 ones is _____ .

- A) 88
- B) 90
- C) 50
- D) 98

4 How many bundles of 10 can Jerry make with his candies?

Jerry can make _____ tens.

PLACE VALUE

4.4 Chapter Review

5 Fill in the missing numbers. 107, _____, _____, 110, _____

6 Rita is thinking of a number that makes both equations true. What number is she thinking of?

_____ < 68   _____ > 66

7 Mayra is counting ice creams on the table. There are 5 new cartons that each have 10 ice cream in them, and 7 more ice cream is available separately on the table. How many ice creams are there on the table?

Ⓐ 52   Ⓑ 57   Ⓒ 38   Ⓓ 42

8 Mike is thinking of a number that is 30 less than 7 tens and 6 ones. What number is he thinking of? How do you know?

_____

PLACE VALUE

Chapter Review 4.4

9 What sign could you put in the circle to compare these two groups?

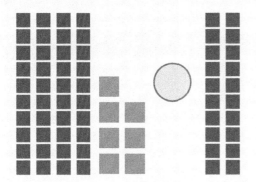

A) =   B) >
C) <   D) +

10 Choose the correct answer from the following.

The sum of 3 tens and 2 ones, 5 tens and 4 ones is _____ .

A) 98   B) 40   C) 86   D) 38

11 How many diamonds?

A) 4 tens 0 ones   B) 5 tens 2 ones
C) 4 tens 6 ones   D) 5 tens 0 ones

PLACE VALUE

4.4 Chapter Review

12 How many tens and ones are in the number ninety-eight?

☐ tens and ☐ ones = 98

13 Fill in the missing numbers.

112, _____ , _____ , 115 , _____

14 How many tens and ones are in the number twenty-nine?

☐ tens and ☐ ones = 29

15 Edwin is counting his birds sticker collection. He put them into Piles of 10. He has 3 piles of 10 and 8 loose birds stickers left over. He gave 2 piles and 7 loose birds stickers to his friend. How many birds stickers Does Edwin have now?

Ⓐ 15   Ⓑ 11   Ⓒ 17   Ⓓ 13

PLACE VALUE

Chapter Review 4.4

16 Count and compare the objects, and then fill in with <, =, and >.

17 What number is four tens more than 4 tens and 8 ones?

_____

18 Choose the correct answer from the following.
The sum of 2 tens and 1 ones ,4 tens and 3 ones is _____ .

Ⓐ 25   Ⓑ 36   Ⓒ 64   Ⓓ 81

19 Fill in the missing numbers.

99, _____ , _____ ,102 , _____

PLACE VALUE

4.4 Chapter Review

20 What number is two ten more than 5 tens and 6 ones?

_____.

Next Chapter: Measurement

CHAPTER 5

MEASUREMENT

DOT-TO-DOT KITTY

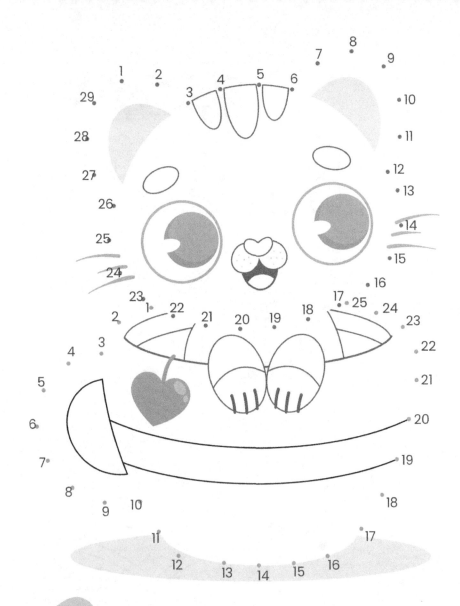

MEASUREMENT

COMPARE AND MEASURE LENGTHS

Comparing lengths means determining whether one object is longer or shorter than another object.

Example 1:

Which tree is the tallest?

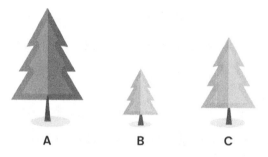

From the above-given trees, we can observe that option (A) tree is the tallest.

Measuring lengths means using standard units of measurement to determine the length of an object.

Example 2:

Measure the length of a pen in inches.

1. Place the pen on a flat surface.
2. Use a ruler to measure the length of the pen from one end to the other.
3. Record the measurement in inches.

MEASUREMENT

5.1 Compare and Measure Lengths

1 Which butterfly is the longest?

 A  B  C  D

2 Angel's marker is 10 inches long, whereas Blessy's marker is 5 inches shorter than Angel's marker, and Mercy's marker is 3 inches longer than Angel's marker. Whose marker is the shortest?

  A) Angel's marker   B) Blessy's marker   C) Mercy's marker

3 Which plant is the tallest?

 A  B  C  D

4 The length of the red stick is 30 cm and the length of the blue stick is 15 cm. Which is the longer one?

_____

MEASUREMENT

Compare and Measure Lengths — 5.1

5 How many dice are needed to measure the rope?

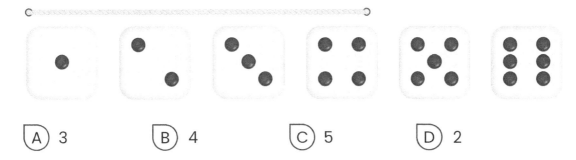

(A) 3   (B) 4   (C) 5   (D) 2

6 Arrange from the longest to the shortest. A carrot of 15 cm, a banana of 30 cm and a radish is 13 cm.

_____

7 How many buttons are needed to measure the length of this line?

(A) 3   (B) 4   (C) 5   (D) 2

8 Arrange from the shortest to the longest. A pen of 12 cm, pencil of 16 cm and ruler is 15 cm.

_____

MEASUREMENT

5.1 Compare and Measure Lengths

9 Mayra caught 3 fish. The first 8 inches long, the second is 6 inches, and the third is 3 inches smaller than the first.

The third fish was _____ inches long.

10 Which one is the smallest one among the following?

Ⓐ   Ⓑ   Ⓒ

11 The length of the pencil is 20 cm, and the length of the glue stick is 10 cm. Which is the longer one?

20cm 10cm

_____

12 Carolina's ribbon is 16 feet long. Andrea's ribbon is 4 feet longer.
True or False: Andrea has a smaller ribbon.

Ⓐ True   Ⓑ False

13 Jack is taller than Julie, but not as tall as Peter. Edward is taller than Peter. Write down the names in order from tallest to the shortest.

_____

MEASUREMENT

Compare and Measure Lengths 5.1

14 True or false:
The banana is 4 buttons long.

Ⓐ True   Ⓑ False

15 There are two cupboards in the kitchen. One of the cupboards is 3 meters long, and the other is 6 meters long. Which cupboard is the smallest?

A   B

_____

16 Arrange from the tallest to the shortest. A pencil of 20 cm, ruler of 30 cm and glue stick is 10 cm.

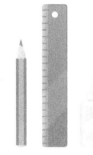

20cm 30cm 10cm

_____

17 Measure the length of the pencil to the nearest centimeter.

_____

MEASUREMENT

5.1 Compare and Measure Lengths

18 Which one is the smallest one among the following?

A   B   C   D

19 Which twig is the shortest?

A
B
C

20 True or False: The line is 3 dice long.

A) True
B) False

MEASUREMENT

CLOCK AND TIME

In a digital clock the time is shown using numbers in which the first number shows the hours and the second number shows the minutes.

An analog clock generally has 2 hands, a short hand and a long hand.. The long hand is known as the minute hand and the short hand is known as the hour hand.

The most commonly used terms regarding time and clock are as shown in the below figure.

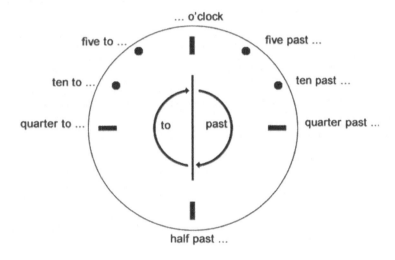

The important terms are described as follows.

"Half past" – means half an hour or 30 minutes.

"Quarter past" – means quarter of an hour or fifteen minutes.

"Quarter to" – means 45 minutes of an hour or 15 minutes more to complete an hour.

MEASUREMENT

CLOCK AND TIME

Example :

1. What time is it when the hour hand is pointing to the 6 and the minute hand is pointing to the 8?

i. Point out the position of both the hour hand and the minute hand on the clock.

ii. Indicate that the hour hand is pointing to the 6 and the minute hand is pointing to the 8.

iii. This means half-past 6 or 6:40.

2. 11:05 means the nearest hour is 11:00 and the nearest half hour for 11:25 is 11:30.

MEASUREMENT

Clock and Time   5.2

1 What time does the clock show?

A) Five thirty B) Six o'clock
C) Six thirty D) Seven thirty

2 What time does the clock show?

A) 1:00 B) 2:00
C) 1:30 D) 2:30

3 True or False:
Half past three is written as 03:30.

A) True B) False

4 What time does the clock show?

A) 2:30 B) 2:00
C) 3:00 D) 3:30

MEASUREMENT

5.2 Clock and Time

5 True or False:
Half past four is written as 04:00.

Ⓐ True   Ⓑ False

6 Which digital clock shows the same time as that of the analog clock?

Ⓐ 10:30   Ⓑ 8:30   Ⓒ 9:30

7 Angel set her alarm to go off so she could get to school on time. Is the clock displaying A.M. or P.M.?

Ⓐ A.M.   Ⓑ P.M.

8 True or False:
Half past nine is written as 08:30.

Ⓐ True   Ⓑ False

9 Edwin is going back to work after he eats lunch at noon. Now the time is 01:00. Is it A.M. or P.M.?

Ⓐ A.M.   Ⓑ P.M.

MEASUREMENT

Clock and Time — 5.2

10 The clocks show when each person got home from the mall. Who got home the latest?

Jenny    Silvia    Blessy

_____ got home the latest.

11 True or False:
Half past nine is written as 01:.00.

Ⓐ True   Ⓑ False

12 What time does the clock show?

_____

13 Tim has to drop his sister, Maria, off at school before he goes to work. Now the time is 08:00. Is it A.M. or P.M.?

Ⓐ A.M.   Ⓑ P.M.

MEASUREMENT

5.2 Clock and Time

14 What time does the clock show?

A) 2:00 B) 3:00
C) 4:00 D) 5:00

15 True or False: Half past 5 is written 5:30.   A) True   B) False

16 True or False: Half past 10 is written 10:30.   A) True   B) False

17 The clock shows 3 things Silvia did on Sunday morning. What did Silvia do first?

Made the breakfast

Clean the car

Workout

Silvia _____ first.

MEASUREMENT

Clock and Time 5.2

18 Choose the correct option from the following. What time is shown on the digital clock?

A. Quarter past 12

B. Quarter to 12

C. 12'o clock

D. Half past 12

19 What time does the clock show?

A. One thirty

B. Six o'clock

C. Twelve o'clock

D. Seven o'clock

MEASUREMENT

5.2 Clock and Time

20 The clocks show the times that Mark and Jack finished their homework on Wednesday evening. Who finished the homework second?

Mark

Jack

_____

MEASUREMENT

Chapter Review 5.3

1 Which pencil is the longest?

Ⓐ

Ⓑ

Ⓒ

Ⓓ

2 What time does the clock show?

Ⓐ 12:30   Ⓑ 11:30

Ⓒ 10:30   Ⓓ 12:00

3 Ria's rope is 12 inches long. Maria's rope is 8 inches shorter than Ria's rope. Silvia's rope is 5 inches longer than Ria's rope. Whose rope is the longest?

Ⓐ Ria's rope   Ⓑ Maria's rope   Ⓒ Silvia's rope

4 What time does the clock show?

Ⓐ Five thirty   Ⓑ Ten thirty
Ⓒ Six thirty   Ⓓ Seven thirty

MEASUREMENT

5.3 Chapter Review

5 How many dice are needed to measure the twig?

- A) 2
- B) 4
- C) 5
- D) 3

6 Which plant is the smallest one among the following?

7 The distance between the point A and point B is 35 km, the distance between the point B to point C is 20 km and the distance between point A and C is 12 km. Which one is the shortest distance?

_____

MEASUREMENT

Chapter Review 5.3

8 What time does the clock show?

- A) Two thirty
- B) Two o'clock
- C) Three thirty
- D) Four o'clock

9 Which one is the larger one among the following?

- A)
- B)
- C)

10 Which digital clock shows the same time as that of the analog clock?

- A) 10:30
- B) 09:30
- C) 10:00

11 Carolina jogged for 7 km whereas Andrea jog 2 km less than Carolina. True or False: Andrea jogged more kilometers than Carolina.

- A) True
- B) False

MEASUREMENT

5.3 Chapter Review

12 True or False:
Half past eleven is written as 11:00.

A) True B) False

13 Steffi is shorter than Julie, but not as short as Mercy. Angel is taller than Julie. Write down the names in order from tallest to the shortest.

_____

14 True or False:
The time shown in the below clock is twelve o'clock.

A) True B) False

15 In a game of hockey, Team A scored 15 points, Team B scored 5 points less than team A and Team C scored 7 points more than Team B. Who scored the maximum points?

A) Team A B) Team B C) Team C

MEASUREMENT

Chapter Review 5.3

16 Choose the correct options from the following. What time is shown in the digital clock?

A) Quarter past 1 B) Quarter to 1
C) 1 o'clock D) Half past 1

17 How many clips are needed to measure the length of this line?

A) 2 B) 4 C) 5 D) 3

18 What time does the clock show?

A) Two o'clock B) Three o'clock
C) Two forty-five D) Seven thirty

MEASUREMENT

5.3 Chapter Review

19 Rita sleeps for 7 hours a day, Mark sleeps of 5 hours a day, and Edwin sleeps for 3 hours more than Mark a day. Who slept for more hours?

(A) Rita (B) Mark (C) Edwin

20 Tim is leaving the school after volleyball practice. Is it A.M. or P.M.?

(A) A.M. (B) P.M.

Next Chapter: Represent and Interpret Data

CHAPTER 6
REPRESENT AND INTERPRET DATA

DOT-TO-DOT BABY DINOSAUR

REPRESENT AND INTERPRET DATA

ORGANIZE AND REPRESENT DATA IN A CHART

Organizing and representing data in a chart is the process of arranging information in simple charts such as pictographs, bar graphs, or tally charts.

Data is a collection of information from facts. The data can be represented in a diagram (or table or chart) when the data is collected and recorded.

Charts or tables, diagrams, or pictures that organize large amounts of data in a clear and concise way. People use charts to interpret current data and make predictions.

Counting the tally of objects, the tally is counted as follows.

| → 1 || → 2 ||| → 3 |||| → 4 |||| → 5

Examples:

The data in a tally chart is clearly understood, and displayed in easy-to-count groups of five. A simple classroom survey about children's favorite pets can be used to reinforce the connection between a tally chart and the information collected. When children see that one tally mark is made for each child's vote, they will understand how to use tally marks to collect and organize data.

Number of dogs are 4 , Number of cats are 5.

REPRESENT AND INTERPRET DATA

6.1 Organize and Represent Data in a Chart

1 Which group has the largest number of items?

A) Zebras
B) Cats
C) Owls
D) Fishes

2 Which group has the fewest number of items?

A) Rings
B) Caps
C) Watches
D) Shoes

3 Make a tally chart of each item.

Apples: _____.

Grapes: _____.

Ice Creams: _____.

REPRESENT AND INTERPRET DATA

Organize and Represent Data in a Chart — 6.1

4 Count and write the tally marks.

Ⓐ |||| Ⓒ ||||| | Ⓑ ||||| Ⓓ ||||| |||

5 Helen made a list of items he wishes to buy from the shop.

Which food item does Helen buy the most of?

Ⓐ Avocados   Ⓑ Apples   Ⓒ Pizzas

6 James made a list of items he wished to buy from the shop.

Which dinner set items does James buy the least of?

Ⓐ Spoons   Ⓑ Forks   Ⓒ Plates

REPRESENT AND INTERPRET DATA

6.1 Organize and Represent Data in a Chart

7 True or false?
The number of cats is greater than the number of dogs.

Ⓐ True　　　Ⓑ False

8 True or false. The number of airplanes is greater than the number of trains?

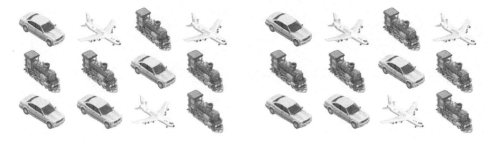

Ⓐ True　　　Ⓑ False

9 Which fruit group has the largest number?

Ⓐ Apples　　　Ⓑ Grapes　　　Ⓒ Strawberries

REPRESENT AND INTERPRET DATA

Organize and Represent Data in a Chart **6.1**

10 State true or false. The number of bees is greater than the number of butterflies.

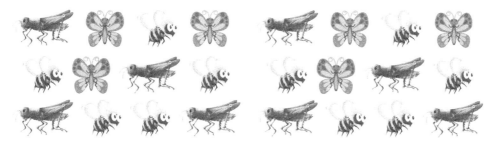

(A) True (B) False

11 Which group has the fewest items?

(A)  (B)

(C)  (D)

12 Which group has the most items?

(A)   (B)

(C)  (D)

13 Which group has the fewest items?

(A)  (B)

(C)  (D)

REPRESENT AND INTERPRET DATA

6.1 Organize and Represent Data in a Chart

14 Which group has the most items?

15 Which tally marks show the correct data?

16 Max looked in the family toy chest and saw some stuffed tortoises and stuffed gold fishes. He made a table of the toys he saw

🐢	🐟
3	6

How many stuffed goldfish did he see?

A) 3   B) 4
C) 6   D) 9

17 Mercy went to a furniture store and saw some chairs and couches. She made a tally chart of the furniture she saw:

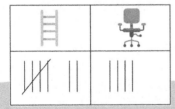

Which piece of furniture did she see fewer of?

A) Ladders   B) Rolling Chairs

REPRESENT AND INTERPRET DATA

Organize and Represent Data in a Chart   **6.1**

18 Make a tally of the number of popsicles.

_____.

19 Make a tally showing the number of guitars.

_____

20 Draw a line from each group of objects to the correct number of tally marks.

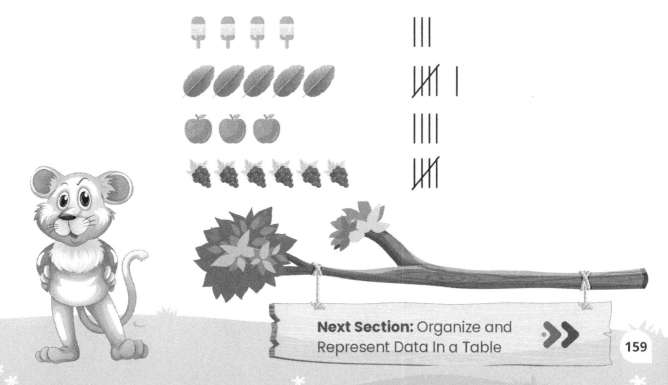

Next Section: Organize and Represent Data In a Table

REPRESENT AND INTERPRET DATA

ORGANIZE AND REPRESENT DATA IN A TABLE

Organizing and representing data in a table is the process of arranging information in simple tables to organize information such as names, numbers, or other data. A table typically consists of rows and columns, with each row representing a separate item or category and each column representing a specific attribute or measurement.

Representing data in the form of a table is known as the Organization of data.

Counting the tally of objects, the tally is counted as follows.

| → 1 || → 2 ||| → 3 |||| → 4 |||| → 5

Example:

Number of toys mentioned in the table

🍍	🐟
3	7

REPRESENT AND INTERPRET DATA

Organize and Represent Data in a Table — **6.2**

1 The teacher conducted a study on the favorite planets of the students and recorded the information in the table.

Planet	Count
Venus	8
Earth	10
Saturn	9

Which planet is the least popular one?

A) Venus
B) Earth
C) Saturn

2 Which planet was the most popular?

Planet	Count
Earth	15
Saturn	12
Jupiter	13

A) Earth
B) Saturn
C) Jupiter

3 The table shows the number of books read by Kerry, Cruz, and Emily on Tuesday and Wednesday.

Names	Tuesday	Wednesday
Kerry	8	9
Cruz	6	8
Emily	7	8

How many books did Emily read on Tuesday and Wednesday?

A) 10
B) 12
C) 14
D) 15

REPRESENT AND INTERPRET DATA

6.2 Organize and Represent Data in a Table

Names	No. of Books
Alan	16
Ava	22
Biju	18
Chris	19

4 Alan, Ava, Biju, and Chris went to the bookstore. Who bought the most books?

(A) Alan (B) Ava
(C) Biju (D) Chris

5 The most rain fell in:

(A) January (B) February
(C) March (D) April

Months	Percentages
January	10%
February	14%
March	12%
April	16%

Names	Candies
Mercy	🍬🍬🍬🍬
Peter	🍬🍬🍬
Polo	🍬🍬🍬🍬🍬
Emma	🍬🍬🍬🍬🍬🍬🍬

6 The chart shows the number of candies that four friends collected during a particular week. How many candies did Polo and Emma collect together?

(A) 10 (B) 12
(C) 14 (D) 9

162

REPRESENT AND INTERPRET DATA

Organize and Represent Data in a Table — 6.2

7. The chart shows the number of candies four friends collected on Halloween. Who collects the fewest candies?

A) Amar
B) Victor
C) Stephy
D) Merlin

Names	Candies
Amar	🎃🎃
Victor	🎃🎃🎃🎃🎃
Stephy	🎃🎃🎃
Merlin	🎃🎃🎃🎃

8. In a donut eating competition, who ate the most?

Names	Donuts
Riya	🍩🍩🍩🍩
Kisha	🍩🍩
Nochu	🍩🍩🍩🍩🍩
Peter	🍩🍩🍩

A) Riya
B) Kisha
C) Nochu
D) Peter

9. Which month has the most holidays?

Months	No. of Holidays
December	10
May	15
January	8
August	5

A) December
B) May
C) January
D) August

REPRESENT AND INTERPRET DATA

6.2 Organize and Represent Data in a Table

Names	Candy
Noah	🍭🍭🍭🍭
James	🍭🍭🍭🍭🍭
Marson	🍭🍭🍭🍭
Jack	🍭🍭🍭

10 True or false? Marson collected the least number of chocolates.

A) True B) False

11 The chart shows the different ways students go to school. How many students preferred bicycles over buses?

Cycle	👤👤👤👤👤
Car	👤👤👤
Bus	👤👤👤👤

A) 1 B) 2
C) 3 D) 4

12 How many cars and bikes are parked in the shopping center parking lot?

A) 7 B) 8 C) 9 D) 10

13 Create a picture graph or a tally chart.

Ice cream	
Apple	
Popcorn	

Organize the data into a table.

REPRESENT AND INTERPRET DATA

Organize and Represent Data in a Table — 6.2

14 Complete the graph or tally chart.

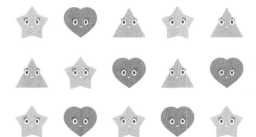

Stars	
Triangle	
Heart	

15 George is a car dealer, and he recorded his sales. State true or false. The number of cars sold by Audi and BMW on Tuesday was the highest.

Car Names	Monday	Tuesday	Wednesday
Audi	67	12	73
BMW	56	28	45

A) True
B) False

16 Use the chart to answer the question.
How many more nuts than nails were left on Friday?

	Wednesday	Thursday	Friday
Bolts	33	20	7
Fasteners	45	18	4
Nails	4	9	4
Nuts	8	5	14

A) 13
B) 12
C) 11
D) 10

REPRESENT AND INTERPRET DATA

6.2 Organize and Represent Data in a Table

17 True or False? John has more shirts than jackets and pants.

Jackets	2
Pants	4
Shirts	8

A) True
B) False

18 How many sundaes did Peter sell on Friday, Saturday, and Sunday.

	Friday	Saturday	Sunday
Fresh Juice	3	2	7
Short Cake	5	1	4
Sundae Cone	4	5	3
Sandwich	8	5	4

A) 11
B) 12
C) 13
D) 14

19 Emma made a list of the flowers in her gardens. Which color Lilly is her favorite?

	Pink	Red	Yellow
Lilly	9	6	8
Rose	12	1	4
Sunflower	2	7	5
Daffodil	3	9	4

A) Pink
B) Red
C) Yellow

REPRESENT AND INTERPRET DATA

Organize and Represent Data in a Table — 6.2

20 The teacher asked the students to vote for their favorite fruit.

Which is the least liked fruit?

Ⓐ Apples   Ⓑ Avocado   Ⓒ Grapes   Ⓓ Pear

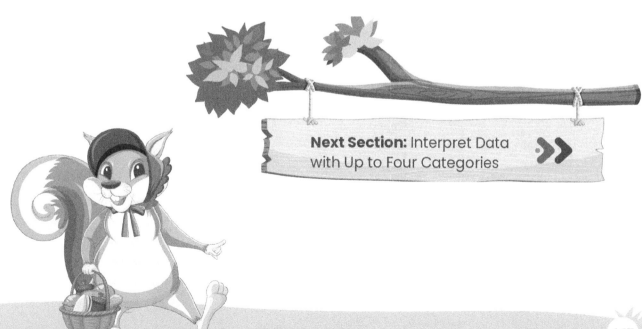

Next Section: Interpret Data with Up to Four Categories

REPRESENT AND INTERPRET DATA

INTERPRET DATA WITH UP TO FOUR CATEGORIES

Interpreting data means making sense of information presented in a visual or numerical format. Different charts, bar graphs, and tables are used to interpret data.

Representing a piece of information from the bar graphs, charts, and tables is known as organizing data.

Examples:

Consider the below example of a survey taken in a class of 24 students about their favorite fruits.

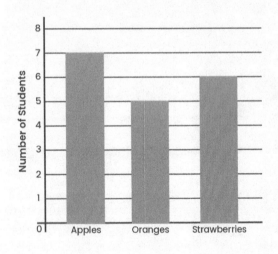

1. Which fruit got the least vote?
Orange
2. Which fruit is the most favorite fruit?
Apples

REPRESENT AND INTERPRET DATA

Interpret Data with Up to Four Categories — 6.3

1 Isabella kept a record of how many pages she read in the past 3 days.

True or False: She read 5 pages on Wednesday.

(A) True (B) False

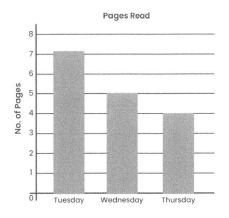

2 Jacob wanted to know which state had the lowest gas prices during the holiday season.

True or False: Colorado has the best gas prices.

(A) True (B) False

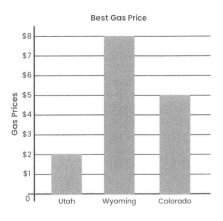

3 Noah and Emma wanted to know how many hours they had worked over the past few days.

True or False: They worked the most hours on Wednesday.

(A) True (B) False

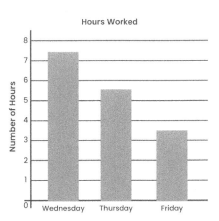

REPRESENT AND INTERPRET DATA

6.3 Interpret Data with Up to Four Categories

4 Sophia went to a furniture store and saw six chairs, seven stools, eight beds, and five lamps. Which bar graph represents the furniture she saw?

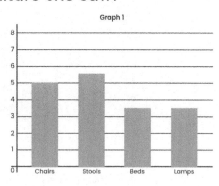
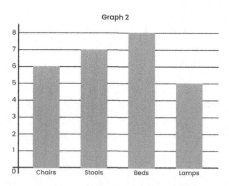

---

5 Lovi went on a picnic and brought Five carrots, Four cupcakes, and six apples. He made a bar graph to show his purchases.
True or False: Graph A is the correct bar graph.

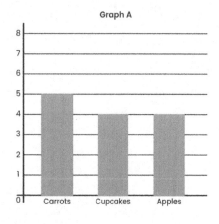

Ⓐ True   Ⓑ False

REPRESENT AND INTERPRET DATA

Interpret Data with Up to Four Categories 6.3

6 How many students like fruit strawberries?

- A) 6
- B) 7
- C) 8
- D) 9

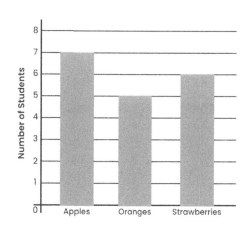

7 Sarah and Pricy removed the leaves from their back yard and they made a chart on the number of leaves collected and their colors. Each leaf corresponds to 5 leaves.

The total number of red and yellow leaves is _____.

Leaf Colors	No. of Leaves
Green	2
Red	4
Yellow	6

8 The shaded blocks show the number of fruits liked by a group of students.

how many prefer grapes to apples?

- A) 1
- B) 2
- C) 3
- D) 4

171

REPRESENT AND INTERPRET DATA

6.3 Interpret Data with Up to Four Categories

9. The below pie chart shows the survey results on their favorite breakfast.
How many more people prefer pancakes than eggs as their breakfast?

A) 40
B) 30
C) 20
D) 10

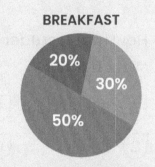

BREAKFAST
20%
30%
50%
■ Toast – 30% ■ Pancake – 50% ■ Egg – 20%

10. Which food item is the most popular?

Strawberries	Cookies and creams	Fudge
⊥⊥⊥⊥ ⊥⊥	⊥⊥⊥⊥	⊥⊥⊥

A) Strawberries
B) Cookies and creams
C) Fudge

11. The graph shows how many vegetables Sally bought at the farm stand. What did Sally buy the least of?

Which is the least preferred vegetable?

A) Cucumber
B) Beet
C) Broccoli

172

REPRESENT AND INTERPRET DATA

Interpret Data with Up to Four Categories 6.3

12 This graph shows the favorite subjects in Mr. Smith's class. How many students like English the best?

What is the total number of votes for English?

Ⓐ 5 Ⓑ 8 Ⓒ 3 Ⓓ 4

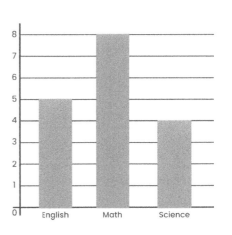

13 A teacher conducted a survey about the favorite candy flavor among a set of students.

Which flavor is liked by exactly six students?

Ⓐ Apple Ⓑ Orange Ⓒ Kiwi

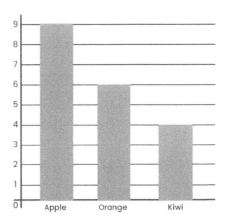

REPRESENT AND INTERPRET DATA

6.3 Interpret Data with Up to Four Categories

14 Use the tally chart and make a bar graph.

Strawberries	Cookies and creams	Fudge
||||	||||/ |||	||||/	

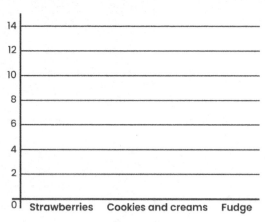

Display the given tally of edibles as colors in the column graph.

15 Which fruit is liked by a large number of students?

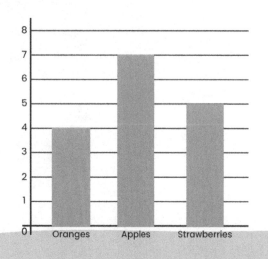

A) Oranges
B) Apples
C) Strawberries

174

REPRESENT AND INTERPRET DATA

Interpret Data with Up to Four Categories   6.3

16 How many students liked the strawberries?

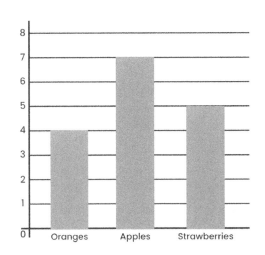

A) 5
B) 10
C) 2
D) 8

17 Which food is liked by exactly seven students?

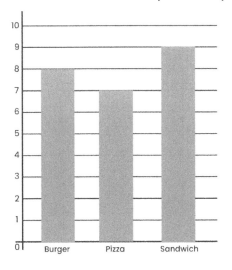

A) Burger
B) Pizza
C) Sandwich

REPRESENT AND INTERPRET DATA

6.3 Interpret Data with Up to Four Categories

18 The graph shows the number of vegetables sold at the farmers market on Wednesday.

What vegetable was sold the most?

A) Cucumber

B) Beet

C) Broccoli

19 Teacher conducted a survey about the favorite candy flavor among a set of students.

Which fruit is liked by 5 students?

A) Bananas   B) Oranges

C) Grapes

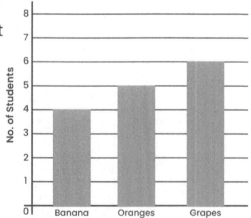

REPRESENT AND INTERPRET DATA

Interpret Data with Up to Four Categories — 6.3

20 Create a bar graph using the data.

Strawberries	Cookies	Cakes
||||/ ||	||||/	||||

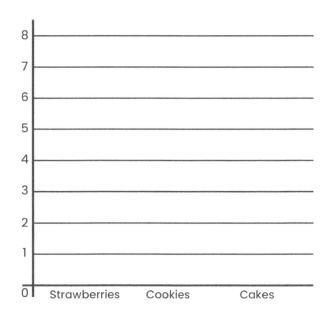

Next Section: Chapter Review

REPRESENT AND INTERPRET DATA

6.4 Chapter Review

1 Which group has the fewest items?

2 Lindo saw a variety of fruit on sale at the farmers' market

Which of the following charts is correct?

A

Fruits	🍎	🥑	🍇	🍐
How many	5	4	6	5

B

Fruits	🍎	🥑	🍇	🍐
How many	5	4	6	4

3 Use the graph to answer the question.
True or False: Mary had 4 trophies

A) True

B) False

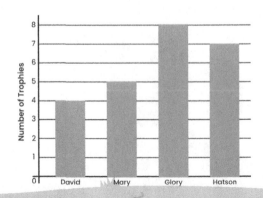

REPRESENT AND INTERPRET DATA

Chapter Review 6.4

4 Sarah is the school photographer and she has to keep a record of how many pictures she takes.

True or False: She took the most pictures on Monday.

A) True B) False

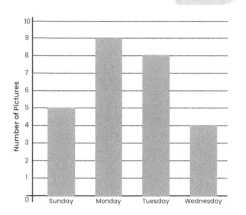

5 Nancy went to the local hardware store and bought ten bulbs, seven flashlights, and five hammers. She made a bar graph of the items he bought.

True or false: Graph A is the correct bar graph.

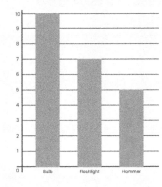 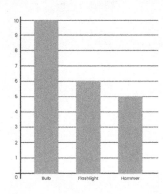

A) True

B) False

6 Fasil sorted through the family junk drawer and found some watches and tape measures. He made a tally chart of the items he found:

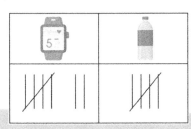

Which item did he find fewer of?

A) Watches

B) Bottles

179

REPRESENT AND INTERPRET DATA

6.4 Chapter Review

7. Mrs. Helen wanted to see how many miles she drove each day so she made a graph to show her mileage. True or False: She drove 10 miles on Sunday.

A) True B) False

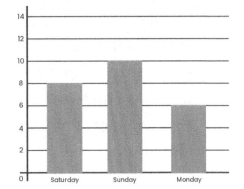

8. Emily and Noah removed the leaves from their backyard and they made a chart on the number of leaves collected and their colors.

Leaf Color	Number of Leaves
Green	2
Red	4
Yellow	6

The total number of green and yellow leaves is _____.

9. In a donut eating competition, Who ate the least?

Names	Donuts
Riya	🍩🍩🍩🍩
Kisha	🍩🍩
Nochu	🍩🍩🍩🍩🍩
Peter	🍩🍩🍩

A) Riya B) Kisha
C) Nochu D) Peter

REPRESENT AND INTERPRET DATA

Chapter Review 6.4

10 Which month had least holidays in a year?

Months	Number of hoildays
December	10
May	15
January	8
August	5

Ⓐ December Ⓑ May
Ⓒ January Ⓓ August

11 The graph represents students based on their mode of travel to school. How did most children come to school?

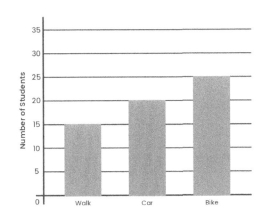

Ⓐ Walk
Ⓑ Car
Ⓒ Bike

12 Is there any equal number of items?

Ⓐ Yes Ⓑ No

REPRESENT AND INTERPRET DATA

6.4 Chapter Review

13 How many hearts?

14 The science teacher conducted a survey about favorite birds. Use the chart to answer the question.

Birds	No. of Students
Parrot	12
Sparrow	14
Duck	20

How many students responded to the survey?

Ⓐ 36 Ⓑ 46
Ⓒ 56 Ⓓ 38

15 The graph shows the number of vegetables sold at the fair on Saturday.

How many fewer cabbages than tomatoes were sold?

Ⓐ 4
Ⓑ 5
Ⓒ 6
Ⓓ 8

REPRESENT AND INTERPRET DATA

Chapter Review 6.4

16. Teacher conducted a survey about the favorite candy flavor among a set of students. Answer the following based on the bar graph given below.

How many students liked the orange?

- A) 10
- B) 6
- C) 12
- D) 8

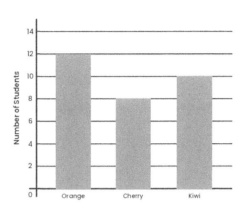

17. How many apples were eaten during snack time in week 1 and week 3?

Week 3

- A) 6
- B) 7
- C) 8
- D) 9

18. Peter counted the stuffed animals in his toy box. He made a table of the toys he saw.

How many stuffed alligators did he see?

- A) Four
- B) Five
- C) Six
- D) Eight

183

REPRESENT AND INTERPRET DATA

6.4 Chapter Review

19 Miss.Chris, a teacher at an elementary school, recorded the favorite subjects of her students in the below chart. Use the chart to answer the question. Which subject is a student's favorite?

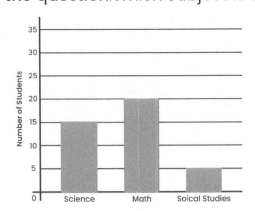

- A) Science
- B) Math
- C) Soical Studies

20 The graph shows the number of toys sold on Friday at Teddy's Toy Shop. Use the graph to answer the question.
How many drum toys were sold?

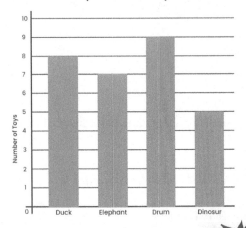

- A) 3
- B) 3
- C) 9
- D) 10

Next Chapter: Geometry

CHAPTER 7
GEOMETRY

DOT-TO-DOT BABY TIGER

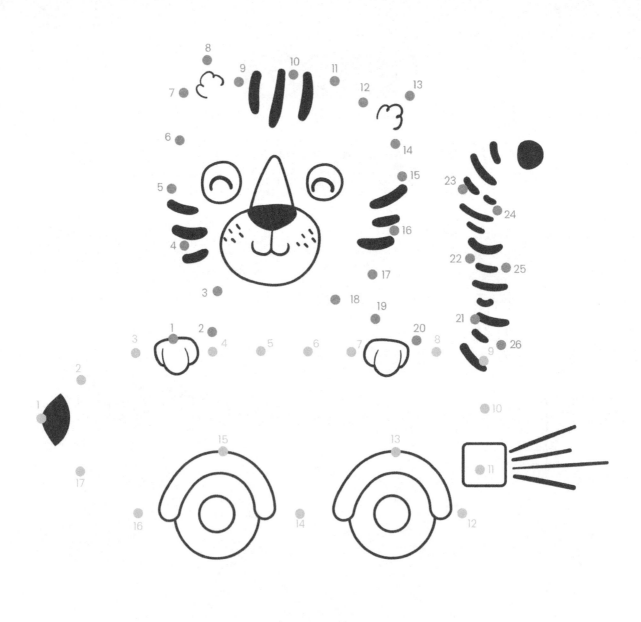

GEOMETRY

GEOMETRIC PROPERTIES - 2D AND 3D SHAPES

2D Shapes properties:

The geometric properties of 2D Shapes are area, perimeter, angles, sides and symmetry.

Area: The amount of space inside a shape.

Perimeter: The length of the boundary of a shape.

Angles: The corner where two lines meet.

Sides: The edges of a shape. For example, a triangle has three sides.

Symmetry: The property of a shape that can be divided into two identical halves. For example, a butterfly has symmetry because both sides are the same.

GEOMETRY

GEOMETRIC PROPERTIES – 2D AND 3D SHAPES

3D Shapes properties:

The geometric properties of 3D Shapes are volume, surface area, height, width, depth, faces, vertices, and edges.

Volume: The amount of space inside a 3D shape.

Surface area: The total area of all the faces of a 3D shape. For example, a rectangular prism has six faces, so its surface area is the sum of the areas of all six faces.

Height: The distance between the top and bottom faces of a 3D shape.

Width: The distance between the left and right faces of a 3D shape.

Depth: The distance between the front and back faces of a 3D shape.

Faces: The flat surfaces that make up the boundary of a 3D shape. For example, a sphere has no flat faces, but a cube has six flat faces.

Vertices: The points where the faces of a 3D shape meet. For example, a cube has eight vertices.

Edges: The line segments that connect the vertices of a 3D shape. For example, a pyramid has edges that connect its vertices to its apex (top point).

An **open shape** is a shape that has at least one side that is not connected to another side, and therefore has an opening. An example of an open shape is a half-circle.

A **closed shape** is a shape where all sides are connected and there are no openings and gaps. An example of a closed shape is a circle.

GEOMETRY

Geometric Properties - 2D and 3D Shapes   7.1

1   Draw a shape using your favorite color.

2   Name a shape with a curved surface.

_____

3   Draw a closed shape with 3 sides.

4   Identify the shape with 0 vertices and 0 sides.

_____

GEOMETRY

7.1 Geometric Properties - 2D and 3D Shapes

5 Is the following shape open or closed?

_____

6 How many sides are there?

_____

7 What 3-D figure has six square faces?

_____

8 Is the following shape open or closed?

Ⓐ Open Ⓑ Close

GEOMETRY

Geometric Properties - 2D and 3D Shapes   **7.1**

9   How many vertices are there in the below shape (ball)?

_____

10   Is the following shape open or closed?

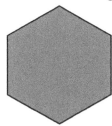

_____

11   How many vertices does a cube have?

A cube has _____ vertices.

12   Identify the shape as having curves and having circular faces on top and bottom.

_____

GEOMETRY

7.1 Geometric Properties - 2D and 3D Shapes

13 What is the name of the shape?

_____

14 Answer the question based on the below shape.
How many sides are there?

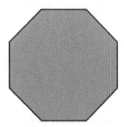   _____

15 Choose the correct answer from the following. Which one of the following shapes is open?

16 Identify a 3-D Figure with 2 faces, 0 edges, and 0 vertices.

_____

GEOMETRY

Geometric Properties - 2D and 3D Shapes — 7.1

17 Draw a shape in two different sizes.

18 Draw two open shapes"?

19 How many edges are there?

_____

20 State true or false. The following shape is a triangle.

_____

Next Section: Identify and Draw 2D and 3D Shapes

GEOMETRY

IDENTIFY AND DRAW 2D AND 3D SHAPES

A 2D shape is a flat plane figure or a shape with two dimensions length and width.

Examples of 2D shapes:

Circle, Rectangle, Triangle, Square etc.

The properties of basic figures are as mentioned below.

- **Circle** - A circle is a 2D shape with no sides.
- **Rectangle** – A rectangle is a 2D shape with four vertices and four sides.
- **Square** – A square is a 2D shape with four vertices and 4 sides with all the sides equal.
- **Triangle** – A triangle is a 2D shape with 3 vertices and sides.

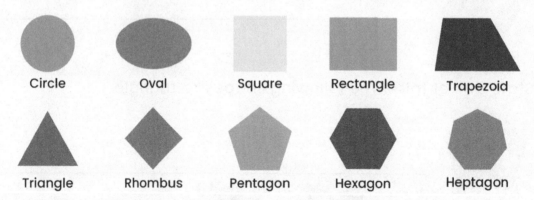

GEOMETRY

IDENTIFY AND DRAW 2D AND 3D SHAPES

A 3D shape is a solid figure or an object with three dimensions which are length, width, and height.

Examples of 3D shapes:

Cube, Rectangular prism, Sphere, Cone, cylinder etc.

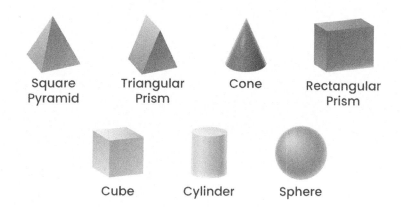

The properties of basic figures are as mentioned below.

Cube – A cube is a 3D shape with 8 vertices, 6 square faces, and 12 edges.

Sphere – A 3D shape with no faces, vertices, and edges.

GEOMETRY

7.2 Identify and Draw 2D and 3D Shapes

1 Draw a triangle.

2 Draw the shape of a diamond.

3 Draw a square inside a circle.

4 Put an x on the circle.

Figure A

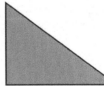
Figure B

GEOMETRY

Identify and Draw 2D and 3D Shapes — 7.2

5 How many sides does a triangle have?

_____

6 Name and trace the figure.

_____

7 How many sides does a heptagon have?

_____

8 Draw the shape of a trapezoid.

GEOMETRY

7.2 Identify and Draw 2D and 3D Shapes

9 Which shape has fewer vertices?

Figure A

Figure B

10 Is the shape open or closed?

Ⓐ Open   Ⓑ Closed

11 Draw a kite.

12 Draw a cylinder.

GEOMETRY

Identify and Draw 2D and 3D Shapes   **7.2**

13   Draw a triangle above a rectangle.

14   True or false?
The square is three-dimensional.   (A) True   (B) False

15   Color the following shapes with a greater number of sides.

Figure A

Figure B

16   Which shape has the least number of vertices?

Figure A   Figure B   Figure C

GEOMETRY

7.2 Identify and Draw 2D and 3D Shapes

17 How many edges are there on a triangular prism?

_____.

18 Which figure has more edges?

Ⓐ Sphere　　Ⓑ Rectangular prism

19 Identify whether the shapes are 2D or 3D.
_____.

20 Identify the name of the shape that is similar to the following figure.
_____.

Next Section: Create Composite Figures in Two or Three Dimensions

GEOMETRY

CREATE COMPOSITE FIGURES IN TWO OR THREE DIMENSIONS

To create a **composite figure in 2D**, combine two or more simpler shapes to create a new shape. For example, a composite figure can be created by combining a square and a triangle to form a house shape.

Example:

Shapes can be placed together to compose a new shape.

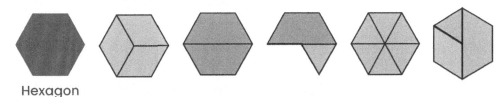

Hexagon

To create a **composite figure in 3D**, combine two or more simpler shapes to form a new 3D shape. For example, a composite figure can be created by combining a cylinder and a cone to form an ice cream cone shape.

Example:

Composing 3D shapes such as cylinder, cube, sphere.

GEOMETRY

7.3 Create Composite Figures in Two or Three Dimensions

1 Which of the following two figures will make a rectangle?

2 Use 4 large squares and make a big square.

3 Use 2 triangles and make the shape of a kite.

4 Draw a shape with four sides and corners such that the length of one side is short and the other side is long.

GEOMETRY

Create Composite Figures in Two or Three Dimensions — 7.3

5 Combine the shapes to make a new one.

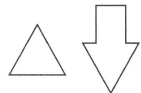

_____

6 Use six triangles to make a hexagon.

7 Sandy was making a shape to be presented to his brother as a gift. He first made a shape with three sides and three corners and then inscribed a shape inside it with no edges or vertices. What will be the shape of the gift?

GEOMETRY

 7.3 Create Composite Figures in Two or Three Dimensions

8 What is the common property in both the figures?

- (A) Open
- (B) 2 Dimensional
- (B) 3 Dimensional

9 Which shape will be formed when a triangle and an inverted triangle are combined?

_____

10 Linda made a shape with 5 edges and 5 corners. What will be the shape?

_____

11 Name the 3D shapes used in the following figure.

_____

GEOMETRY

Create Composite Figures in Two or Three Dimensions — 7.3

12. Whether the statement is correct or incorrect When two squares are joined, they form a square.

Ⓐ Correct   Ⓑ Incorrect

13. Which figures are there in the following figure?

_____

14. Which 3D shape are there in the following figure?

_____

15. What 3D shape is similar to a carrot?

_____

GEOMETRY

7.3 Create Composite Figures in Two or Three Dimensions

16 Which all 3D shapes are there in the following figure?

_____

17 Which 3D shapes are used in the following figure?

_____

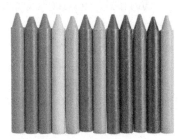

18 Name 2 shapes you can make by splitting the following square into two equal parts.

_____

19 Ava made a shape with 3 edges and 3 corners. What will be the shape?

_____

GEOMETRY

Create Composite Figures in Two or Three Dimensions — 7.3

20 Does this figure have a curved surface?

Ⓐ Yes Ⓑ No

Next Section: Chapter Review

GEOMETRY

7.4 Chapter Review

1 Which two shapes are used in making the following figure?

_____

2 Samuel takes a square-shaped piece of paper. He then folds diagonally by joining opposite corners. What two-dimensional shape is the new folded piece of paper?

Ⓐ Square Ⓑ Circle Ⓒ Rhombus Ⓓ Triangle

3 Tim is buying a wall poster. The wall poster has 4 corners. All the sides are same length. What shape is the wall poster?

Ⓐ Square Ⓑ Rectangle Ⓒ Trapezoid Ⓓ Triangle

4 Which 2D shapes are used to make the following figure?

_____

GEOMETRY

Chapter Review 7.4

5 Robert drew a shape with 0 corners and 0 straight sides. What shape did he draw?

Ⓐ Diamond Ⓑ Rhombus Ⓒ Circle Ⓓ Rectangle

6 Name a shape you can make by splitting the following square diagonally twice.

Ⓐ Square Ⓑ Rectangle
Ⓒ Trapezoid Ⓓ Triangle

7 Patty bought a basketball. When she got home, she took the ball out of a cube-shaped box. What shape is the basketball?

Ⓐ Square Ⓑ Rectangle Ⓒ Trapezoid Ⓓ Sphere

8 As Markie walked to school, she noticed the crossing sign across the street. It had four equal sides and four corners. It looked like a diamond. What shape is the crossing sign?

Ⓐ Square Ⓑ Rectangle Ⓒ Rhombus Ⓓ Sphere

GEOMETRY

7.4 Chapter Review

9 How many vertices does the following shape have?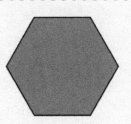

_____

10 Draw a shape with pentagon.

11 Is this 2D or 3D figure?

_____

12 How many sides does the following shape have?

_____

GEOMETRY

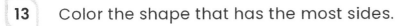

Chapter Review 7.4

13 Color the shape that has the most sides.

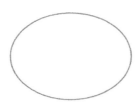 

14 True or false?
The Pentagon is three-dimensional.   Ⓐ True   Ⓑ False

15 Is the shape open or closed?

   Ⓐ Open   Ⓑ Closed

16 Identify a shape with 3 sides and 3 vertices.

_____

17 Draw a closed shape with 5 sides.

211

GEOMETRY

7.4 Chapter review

18 How many vertices does this shape have?

_____

19 Determine whether the following figure has a curved surface.

Ⓐ Yes   Ⓑ No

20 Which 2D shapes are used in the following figure?

_____

Next Chapter: Partition Shapes

CHAPTER 8

PARTITION SHAPES

DOT-TO-DOT BUTTERFLY

PARTITION SHAPES

DIVIDE CIRCLES AND RECTANGLES INTO TWO OR FOUR EQUAL PARTS

Partitioning refers to dividing the given object into two or more parts.

If a shape is divided into two equal parts, then each share is called as a half and the shape is then called as halves. The following figures are examples of dividing the shapes of circles and rectangles into two equal halves.

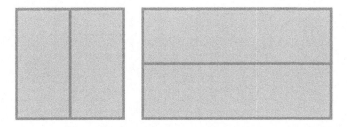

If a shape is divided into four equal parts then each share is called as a quarter and the shape is then called as fourths. The following figures are examples of partitioning are the shapes of circles and rectangles into 4 equal halves.

 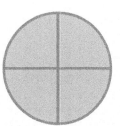

PARTITION SHAPES

 8.1 Divide Circles and Rectangles into Two or Four Equal Parts

1 Select the picture that show equal parts.

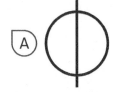   

2 Select the picture that shows equal parts.

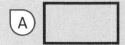  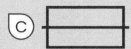 

3 Which figure shows halves?

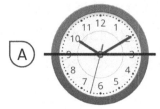 

4 Which figure shows fourths?

  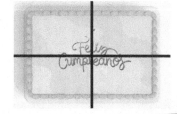

PARTITION SHAPES

Divide Circles and Rectangles into Two or Four Equal Parts 8.1

5 Which choice shows fourths?

A 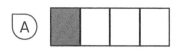   B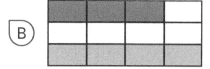

C 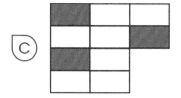   D

6 State whether it is true or false:

The given watermelon is cut equally in half.

A) True   B) False

7 Partition the circle into two equal halves

8 The following diagram is divided into _____ parts.

PARTITION SHAPES

 8.1 Divide Circles and Rectangles into Two or Four Equal Parts

9 Partition the lemon into two equal halves.

10 Partition the circle into two parts.

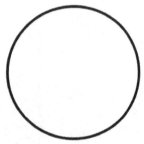

11 Shade half of the shape.

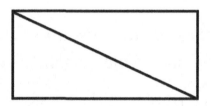

12 True or false.
The following shape is divided into four equal halves.

_____

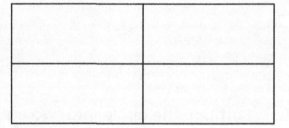

13 True or false. The following shape is divided into four equal halves.

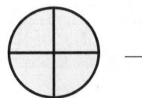 _____

PARTITION SHAPES

Divide Circles and Rectangles into Two or Four Equal Parts 8.1

14 True or false. The following picture is divided into 2 equal parts.

_____

15 True or false. The following picture is divided 2 equal parts.

_____

16 Partition the pizza into 2 halves.

17 True or false.
The following eraser picture is divided into 2 equal parts.

_____

PARTITION SHAPES

 8.1 Divide Circles and Rectangles into Two or Four Equal Parts

18 Partition the wheel into four equal parts?

19 The given circle is divided into how many parts?

A) 2   B) 4
C) 6   D) 8

20 The given dartboard is divided into how many parts?

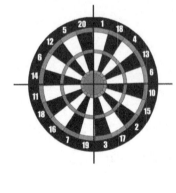

Next Section: Describe Shares as Halves, Fourths, and Quarters

PARTITION SHAPES

DESCRIBE SHARES AS HALVES, FOURTHS, AND QUARTERS

The word half means the shape is divided into two equal halves where each part is known as a half. The word quarter means the shape is divided into four parts or fourths where each part is known as a quarter.

Examples:

On dividing the whole or a group of objects is into equal parts, we get equal shares. We need to divide an object or a whole number into equal parts to distribute it equally.

If a whole is divided into four equal parts, each part is a FOURTH. A fourth is obtained by dividing a whole (1) by 4.

221

PARTITION SHAPES

 8.2 Describe Shares as Halves, Fourths, and Quarters

1 Is the shape divided into halves?  _____

2 Is the given shape divided into fourths?  _____

3 Divide the following shape into halves? _____

4 Which following shape is divided into fourths?

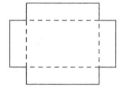 _____

5 Color the half of the shape.

PARTITION SHAPES

Describe Shares as Halves, Fourths, and Quarters — 8.2

6 True or false? The following shape is divided into halves

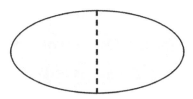 _____

7 What part of the figure is shaded?

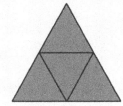

- Ⓐ One-half
- Ⓑ Whole shape
- Ⓒ One-quarter
- Ⓓ No part is shaded

8 The given shape is divided into how many parts?

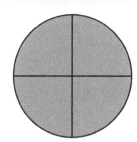

_____

9 Match the following

   One half shaded

   No part shaded

   One quarter shaded

   Whole shape shaded

PARTITION SHAPES

8.2 Describe Shares as Halves, Fourths, and Quarters

10 How many parts are colored?

A) One half
B) Whole shape
C) One quarter
D) No part is shaded

11 How many parts of the below shape are unshaded?

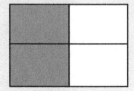

_____

12 What part is shaded?

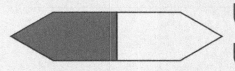

_____

13 Which shape is divided into less equal parts?

 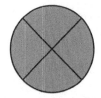

_____

PARTITION SHAPES

Describe Shares as Halves, Fourths, and Quarters 8.2

14 How many parts are shaded?

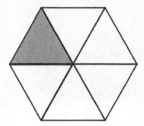

_____

15 Is this partitioned into halves or quarters?

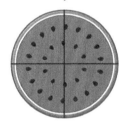

_____

16 Is the partitioned in halves or fourths?

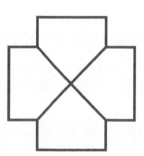

_____

17 How many parts?

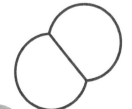

_____

_____

PARTITION SHAPES

8.2 Describe Shares as Halves, Fourths, and Quarters

18 Identify whether the below shape is divided into halves or quarters?

_____

19 The given shape is divided into how many parts?

_____

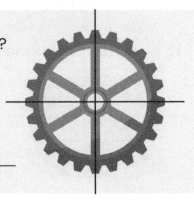

20 Is the shape divided into halves?

_____

Next Section: Identify Equal and Unequal Shares

PARTITION SHAPES

IDENTIFY EQUAL AND UNEQUAL SHARES

When objects are divided into equal parts, we call them equal shares. Equal shares can be equal halves, equal quarters, Equal thirds, etc.

When objects are divided into unequal parts, we call them unequal shares.

Example:

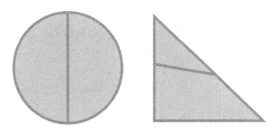

Shapes A is divided into equal shares.

Shapes B is divided into unequal shares.

PARTITION SHAPES

8.3 Identify Equal and Unequal Shares

1 Identify the equal shares in the below diagram. (circle the correct answer)

2 Identify the unequal shares from the below diagram. (circle the correct answer)

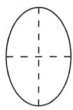 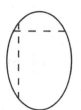

3 Which shape shows equal shares? (circle the correct answer)

4 Choose the correct answers.

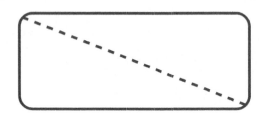

A) Equal shares

B) Unequal shares

PARTITION SHAPES

Identify Equal and Unequal Shares 8.3

5. Are the below pictures shared equally or unequally?

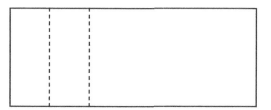

6. Are the below pictures shared equally or unequally?

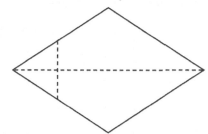

A) Yes

B) No

7. Which one of the following shows equal shares?

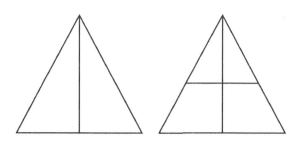

PARTITION SHAPES

8.3 Identify Equal and Unequal Shares

8 Mark the picture to show equal shares.

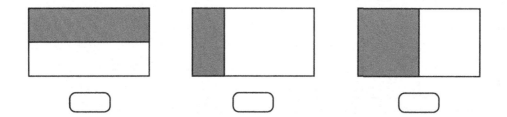

9 Answer yes or no. It is possible to divide the given shape into equal shares.

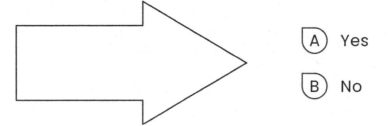

A) Yes

B) No

10 Divide the shape of into 2 equal parts

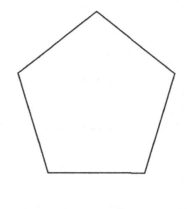

PARTITION SHAPES

Identify Equal and Unequal Shares 8.3

11 Divide the shapes into two equal shares.

12 Answer yes or no. It is possible to divide the given shape into equal shares.

Ⓐ Yes

Ⓑ No

13 Mark the picture to show equal shares.

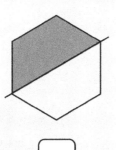 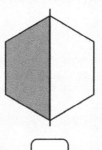 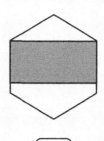

PARTITION SHAPES

8.3 Identify Equal and Unequal Shares

14 Mark the picture to show unequal shares.

  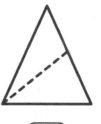

☐   ☐   ☐

15 Select the correct option

- **A** The given shape is divided into equal shares.
- **B** The given shape is divided into unequal shares.
- **C** The given shape is divided into quarters.
- **D** The given shape is a whole.

PARTITION SHAPES

Identify Equal and Unequal Shares 8.3

16. Jack has a donut. Is it possible for him to divide the set into equal halves?

17. Mary has an apple. Is it possible to divide the apple into equal halves?

18. Mark the picture to show unequal shares.

233

PARTITION SHAPES

8.3 Identify Equal and Unequal Shares

19 Mark the picture to show equal shares

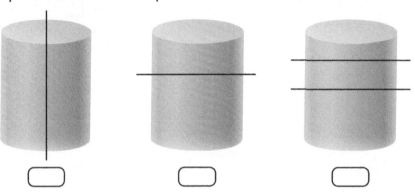

20 Select the correct options from the following

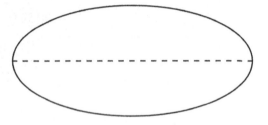

Ⓐ Equal shares   Ⓑ Unequal shares

PARTITION SHAPES

Chapter Review 8.4

1 Which figure shows halves?

 (A)  (B)

2 Which figure shows fourths?

(A)    (B)

3 How the rectangle is divided?

(A) Equal parts

(B) Unequal parts

PARTITION SHAPES

 8.4 Chapter Review

4 The following diagram is divided into _____ parts.

A) Zero B) Two parts
C) Three parts D) Four parts

5 The following diagram is divided into _____ parts.

A) 1 B) 0
C) 2 D) 3

6 Which picture shows the equal shares?

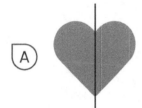 

A B

PARTITION SHAPES

Chapter Review — 8.4

7 What part of the shape is shaded?

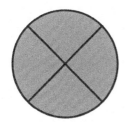

- A) Quarters
- B) Halves
- C) Thirds
- D) Whole

8 Which figure shows fourths?

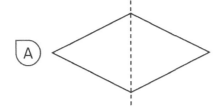 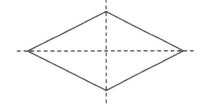

9 **True or false:** The given candy is cut equally in half.

- A) True
- B) False

10 The following diagram is divided into _____ parts.

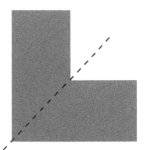

PARTITION SHAPES

8.4 Chapter Review

11 Partition the tomato into two equal halves.

12 Shade half of the shape.

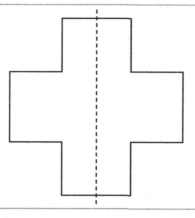

13 Divide the rectangle into 4 equal parts.

14 Is the shape divided into halves?

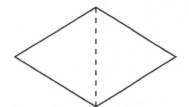

Ⓐ Yes   Ⓑ No

238

PARTITION SHAPES

Chapter Review 8.4

15 The given shape is divided into how many parts?

_____

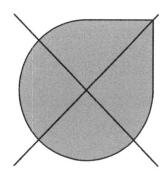

16 How many parts of the given shape are colored?

A) One half   B) Whole shape
C) One quarter   D) No part is shaded

17 Identify the below picture is divided into halves or quarters

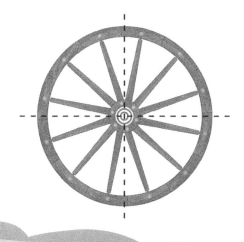

239

PARTITION SHAPES

8.4 Chapter Review

18 Is the given shape divided into fourths?

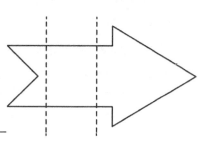

_____

19 Divide the following shape into quarters?

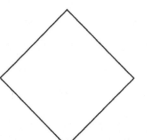

20 What part of the figure is shaded?

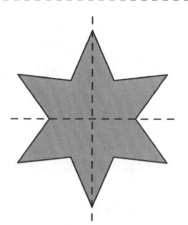

- (A) One half
- (B) Whole shape
- (C) One quarter
- (D) No part is shaded

ASSESSMENT 1
COMPREHENSIVE ASSESSMENTS

DOT-TO-DOT LION

ASSESSMENT – 1

1 Rita had 35 heart stickers. Her friend gave her 20 more stickers. How many stickers does she have altogether?

(A) 55   (B) 45   (C) 65   (D) 35

2 Which number goes in the box to make the equation true?

(A) 4   (B) 6
(C) 7   (D) 5

3 Tim has 5 bikes, Jessica has 9 bikes, and Riya has 6 bikes. How many do do they have altogether?

(A) 22   (B) 18   (C) 20   (D) 15

4 Use the picture to complete the equation.

15 = _____ + 6

5 David has football practice 12 days this month and 7 days next month. How many days of practice does he have in total?

(A) 22   (B) 19   (C) 20   (D) 15

ASSESSMENT – 1

6 To simplify the equation " 7 + 3 + 6 = ?" into the equation "10 + 6 = ?", what do you need to do?

- (A) Add 7 and 3
- (B) Add 3 and 6
- (C) Add 7 and 6
- (D) Subtract 1 from 7

7 Sarah completed 7 flower paintings and Benny completed 4 flower paintings. What is the total number of paintings completed by them?

_____

8 There are 8 red roses, 5 pink roses, and 4 yellow roses in the garden. Which equation shows how many roses are there in the garden and find their total?

- (A) ? + 4 + 8 = ?
- (B) 8 − 5 + 4 = ?
- (C) 8 + 5 + 4 = ?
- (D) 4 + 7 + 2 = ?

9 Use the picture to help you solve the equation below.

14 - _____ = 9

10 Elisa bought 7 apples and 14 strawberries from the shop. How many more strawberries does Elisa have than apples?

- (A) 8
- (B) 4
- (C) 7
- (D) 6

ASSESSMENT – 1

11 Max went to a birthday party. There were 13 cupcakes at the party, and 7 cupcakes were eaten. How many cupcakes were left? He made this drawing, to help him solve the problem:

What should he do next?

- A) Draw 3 more cupcakes.
- B) Draw 1 more cupcake.
- C) Cross out 16 cupcakes.
- D) Cross out 7 cupcakes.

12 Fill in the blanks.

Franklin brought 6 burgers from the shop. His mom gave him _____ burgers. Now he has 13 burgers.

13 Tom bought 12 apples. He gave 4 apples to his friend. He wanted to give the remaining apples to his sister. How many apples did he give his sister?

- A) 3
- B) 4
- C) 5
- D) 8

14 Count the blocks and choose the correct answer.

- A) 6
- B) 8
- C) 9
- D) 5

245

ASSESSMENT – 1

15 Steffi bought 9 golden and 5 silver dresses for the event. How many dresses has she bought altogether? Write to equations you could use to solve this problem.

16 True or false: This shape shows halves.

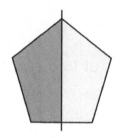

Ⓐ True　　Ⓑ False

17 Joe says the expressions 18 − 3 and 19 − 4 are not equal. Tim says they are equal. Who is right?

Ⓐ Joe　　Ⓑ Tim

18 Write the expression that matches the groups of vegetables on each box. If they have the same amount of vegetables. Write the equal sign between the expressions.

_____ + _____ ☐ _____ + _____

246

ASSESSMENT – 1

19 Solve the sentences and write the 10 + or 10– fact.

In a party, there are 8 burgers, 5 hotdogs, and 2 pizzas. How many items are there in the party? 10 + _____ = _____ .

20 Joseph has 8 red kites and 6 blue kites. Then, 3 kites flew away. How many kites are left with Joseph?

21 Choose the correct answer.

Angel decided to count up to one hundred and fifteen. She counted up to one hundred and eight and got stuck. What number should she say next?

(A) 116 (B) 120 (C) 109 (D) 118

22 Masha completed reading 15 questions and Ronald completed 25 questions for the test. They wrote the following equation to help them compare the questions completed: 15 < 25. Do you agree or disagree? Why?

_____

23 How many tens and ones are in eighty-nine?

(A) 8 tens 9 ones (B) 8 tens 5 ones

(C) 8 tens 2 ones (D) 8 tens 7 ones

ASSESSMENT – 1

24. Jessica has 3 sticks of tens cubes. She gets a new pack that has 9 ones cubes. How many cubes does she have in all?

- (A) 36
- (B) 39
- (C) 34
- (D) 32

25. Silvia has counted 37 green apples and 23 mangoes in her fruit shop. She wants to add tens and some ones to find out how many fruit she has seen in all. What equation could she use to help her?

- (A) 30 + 7 + 23 + 3
- (B) 30 + 5 + 20 + 9
- (C) 30 + 5 + 20 + 0
- (D) 30 + 7 + 20 + 3

26. Ben has 3 note books. He counts 37 empty pages in the first notebook. 10 empty pages in the second notebook, and 12 empty pages in the third notebook. How many empty pages are there in the three books?

- (A) 57
- (B) 62
- (C) 48
- (D) 59

27. How many bundles of 10 can Jessy make with her flowers?

Jenny can make _____ tens.

ASSESSMENT – 1

28 What time does the clock show?

- A) 1:30
- B) 3:30
- C) 02:30
- D) 2:10

29 Write the number as tens and ones.

5 tens and 9 ones – 2 tens and 7 ones = _____ tens and _____ ones.

30 How many dice are needed to measure the length of the line?

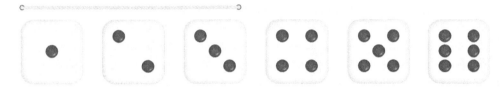

- A) 4
- B) 3
- C) 5
- D) 2

31 Catherina's classroom is 15 feet long. Bethel's classroom is 3 feet longer than Catherina's.

True or False: Bethel has a shorter classroom.

- A) True
- B) False

32 The time shown in the below clock is 30 minutes slow. Write down the correct time.

_____.

ASSESSMENT – 1

33 Use a number line to find the answer.

8 + 2 = _____ .

0 1 2 3 4 5 6 7 8 9 10 11 12 13

34 Which number goes in the blank to make the equation true?
_____ + 6 = 18.

- (A) 11
- (B) 12
- (C) 15
- (D) 13

35 What fraction of the shape is not shaded?

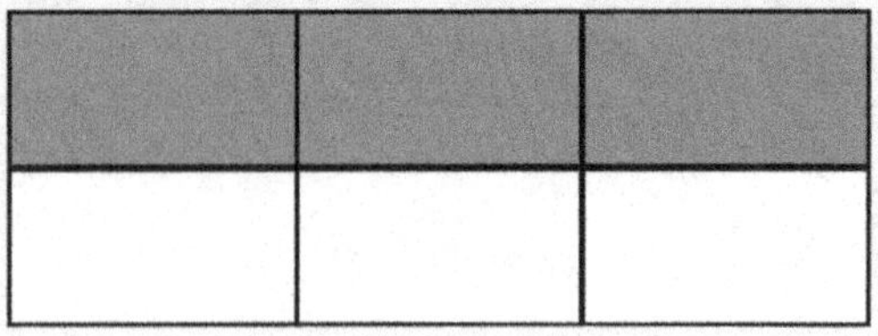

- (A) $\frac{3}{6}$
- (B) $\frac{2}{6}$
- (C) $\frac{2}{8}$
- (D) $\frac{3}{3}$

36 Add the two numbers and write the sum.

5 + 4 = _____ .

37 Ava has 9 cupcakes. Her friend gives 6 more cupcakes. How many cupcakes does she have?

- (A) 16
- (B) 13
- (C) 11
- (D) 15

ASSESSMENT – 1

38 Which number goes in the blank? Write 10+ strategies.
3 + 3 + 8 = 10 + _____.

39 John had 10 sketches. He gave some to his brother. Now he has 5. How many sketches did John give his brother?
10 - _____ = 5

(A) 8 (B) 5 (C) 9 (D) 7

40 What kind of food is the most popular?

Burgers	Pizzas	Hotdogs
𝍤 ‖	‖‖‖‖	‖‖‖

(A) Burgers
(B) Pizzas
(C) Hotdogs

41 What is the common property in both figures?

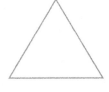 

_____

42 Teena is buying a flower painting. The painting has 4 corners and 4 sides. Two sides are long, two sides are short. What shape is the painting?

(A) Square (B) Rectangle (C) Circle (D) Triangle

ASSESSMENT – 1

43 Jenifer bought 24 oranges and her mom gave her 7 more. How many oranges does Jenifer have in total?

(A) 39 (B) 29 (C) 31 (D) 35

44 Which shape has the most vertices?

(A) (B) (C)

45 Jessy is preparing breakfast for her family. The time on the clock shows 8:30.
True or false: Is it A.M.

(A) True (B) False

DOT-TO-DOT PARROT

ASSESSMENT – 2

1 Alice bought 8 ice cream treats from the shop. Her friend gave 9 more ice creams. How many ice cream treats does she have in all?

(A) 17 (B) 11 (C) 12 (D) 15

2 Use the picture to solve the equation 11 + 9 = ?

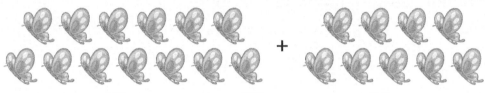

(A) 17 (B) 19 (C) 20 (D) 15

3 Mercy will draw 7 pictures on Monday, 4 pictures on Tuesday, and 3 pictures on Wednesday. How many pictures will Mercy draw altogether?

(A) 17 (B) 14 (C) 20 (D) 15

4 Which number goes in the blank to make the equation true?

5 + 9 = _____ + 6

5 Tim has 7 apples. His mom gave him 5 strawberries and 3 oranges. How many fruits does he have now? Write 10+ strategies.

_____ Apples + _____ Oranges + _____ Strawberries =

_____ fruits in total

ASSESSMENT – 2

6 Lisa bought 7 photo frames from the shop. Ellen bought 3 more frames than Lisa. How many frames did they have altogether?

 A) 15   B) 19   C) 20   D) 17

7 Lisa bought 29 chocolates to distribute to her friends. She distributed 19 chocolates. How many chocolates does she have left?

 A) 10   B) 9   C) 7   D) 11

8 Rita has 7 red marbles, 5 blue marbles, and 4 pink marbles. How many marbles does she have?

_____

9 Count and write the tally marks. _____ Tally Marks

 A) 8   B) 7   C) 9   D) 10

10 Which subtraction equation could represent the picture?

 A) 9 + 3 =?   B) 3 − 9=?   C) 10 − 4=?   D) 14 − 4=?

ASSESSMENT – 2

11 There are 15 vegetables in the basket. Four vegetables were tomatoes, five were carrots, and the rest were peppers. How many peppers were there?

(A) 7 (B) 6 (C) 8 (D) 9

12 David is solving the equation "14 − 6 = ?".
First, he rewrote the equation as "14 − 4 − 2 = ?" Then,
He rewrote the equation as "10 − _____ = ?".
What number goes in the blank? How do you know?

13 Andrew has 15 toys. He gave 8 toys to his friend. How many toys does Andrew have left?

(A) 7 (B) 6 (C) 8 (D) 9

14 Use the picture to help you solve the equation below.

12 − _____ = 7

15 Choose the best answer: 7 + 5 = _____ .

(A) 8 + 5 = _____   (B) 7 + 2 = _____

(C) _____ = 4 + 9   (D) _____ = 5 + 7

ASSESSMENT – 2

16. Angel bought 3 lilies, 7 roses, and 4 lotuses from the flower shop. How many flowers does she have altogether?

☐ + ☐ + ☐ = ☐

10 + ☐ = ☐

17. David bought 5 black and 6 blue jeans for the dance program. How many jeans did he buy altogether?

Ⓐ 14   Ⓑ 13   Ⓒ 11   Ⓓ 12

18. Which equation is equal to 9 + 5 = _____?

Ⓐ 8 + 6 = _____.   Ⓑ 3 + 4 = _____.

Ⓒ 8 + 3 = _____.   Ⓓ 2 + 9 = _____.

19. Which subtraction fact is in the same fact family as 7 + 7 = _____.

Ⓐ 14 – 7   Ⓑ 9 – 2   Ⓒ 11 – 3   Ⓓ 17 – 3

20. True or false?

Ⓐ True   Ⓑ False

ASSESSMENT – 2

21 Rita bought 8 sandwiches, 5 hotdogs, and 3 spring rolls from the shop. How many items does she have now?

☐ + ☐ + ☐ = ☐

22 Count and compare the objects and then fill with <, =, and >.

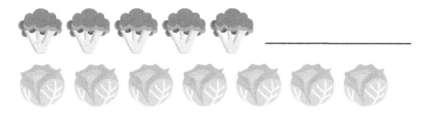

_____

23 Use <, > or, = to complete the inequality.

8 tens 5 ones _____ 5 tens 9 ones.

24 There are 29 drawings in the drawing book. Lisa has already finished 10 drawings. How many drawings does she have to do?

Lisa has _____ drawings to do.

25 Jessica is counting by ones. What number should she say after eighty-two?

(A) Seventy-nine (B) Eighty-three

(C) Eighty-nine (D) Eighty-four

ASSESSMENT – 2

26. What is 98 minus 4 tens and 5 ones?
_____ tens and _____ ones = _____ .

- A) 53
- B) 43
- C) 36
- D) 58

27. How many tens and ones are in the number ninety-three? Explain how you know.

_____

28. How many tens and ones are shown?

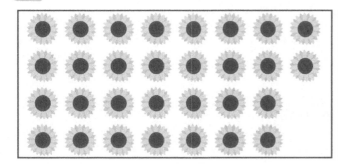 

- A) 4 tens 2 ones.
- B) 5 tens 2 ones
- C) 4 tens 6 ones
- D) 4 tens 0 ones

29. Ellen has 50 cubes. She gave 2 tens to her friend. How many cubes will she have?

- A) 60
- B) 30
- C) 40
- D) 73

30. What sign could you put in the blank to compare these two groups?

- A) =
- B) <
- C) >
- D) +

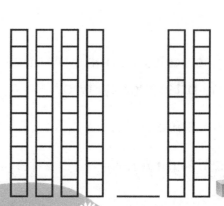

ASSESSMENT – 2

31 There are two shelves in the bedroom. One of the shelves is 4 meters long and other is 7 meters long. Which shelf is the longest?

_____

32 What time does the clock show?

A) Nine thirty B) Nine o'clock

C) Six thirty D) Seven thirty

`09:30`

33 Fill in the blanks with 10 less or 10 more.

_____ , 58 , 68, _____ .

34 Arrange the items from the largest to the smallest. A red rope is 12 cm, a blue rope 30 cm and a black rope is 22 cm.

_____

35 How many items are not shaded?

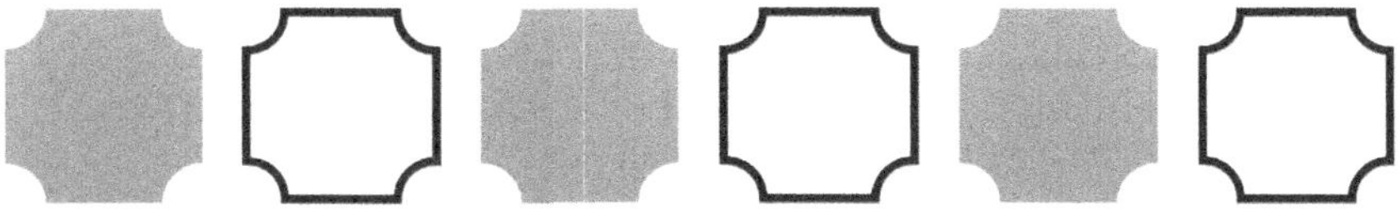

ASSESSMENT – 2

36. Noah looked in the family toy chest and saw some stuffed tortoises and stuffed gold fishes. He made a table of the toys he saw

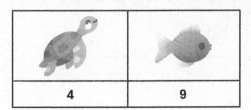

How many stuffed goldfish did he see?

A) 3   B) 4
C) 6   D) 9

37. Heena made a list of items he wishes to buy from the shop.

Food Items	Avocados	Apples	Pizzas
How many	7	5	4

Which food item Heena buys the most of?

A) Avocados   B) Apples   C) Pizzas

38. State true or false. The number of cars is greater than the number of trains?

A) True

B) False

39. Which food item is the most popular?

Strawberries	Cookies and creams	Fudge
|||| /	||||	|||

A) Strawberries

B) Cookies and creams

C) Fudge

ASSESSMENT – 2

40 The chart shows the number of stickers collected by four friends in a particular week.

How many stickers did Steffi and Angel get together?

Names	Stickers
Steffi	☺ ☺ ☺ ☺ ☺ ☺ ☺ ☺
Elisa	☺ ☺ ☺ ☺ ☺
Angel	☺ ☺ ☺
Elizabeth	☺ ☺ ☺ ☺

(A) 12 (B) 10
(C) 9   (D) 7

41 How many students like the Broccoli?

(A) 11 (B) 10
(C) 9   (D) 8

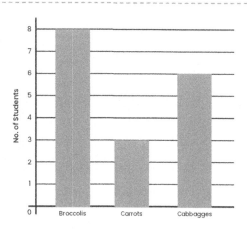

42 The colored blocks show the number of vegetables liked by a group of people.

Which is the most popular vegetable?

(A) Cucumbers
(B) Beets
(C) Broccoli

ASSESSMENT – 2

43 Linda made a shape with 6 edges and corners.
What will be the shape?

_____

44 How many buttons are needed to measure the length of this line?

(A) 3 (B) 4 (C) 5 (D) 2

45 James is showing the number 28 in the base 10 with his blocks. He already has these blocks. What else does she need?

(A) 10 ten sticks (B) 2 ten sticks

(C) 4 ones (D) 11 ones

Next Chapter: Answers and Explanations

ANSWERS AND EXPLANATIONS

TABLE OF CONTENTS

CHAPTER 1 ADDITION WITHIN 20

1.1 Addition Basics 268
1.2 Finding Missing Addend 269
1.3 Three Number Addition With Counting on and Making Ten 270
1.4 Addition Word Problems 271
1.5 Chapter Review 272

CHAPTER 2 SUBTRACTION WITHIN 20

2.1 Subtraction Basics 274
2.2 Subtraction by Decomposing a Number into Ten 275
2.3 Subtraction Word Problems 276
2.4 Chapter Review 277

CHAPTER 3 PROPERTIES OF ADDITIVE AND SUBTRACTIVE OPERATIONS AND EQUATIONS

3.1 Commutative Property and Associative Property of Addition 279
3.2 Three Number Addition or Subtraction 280
3.3 Add or Subtract Within 100 Using Properties of Operations 281
3.4 Finding Two Values Are Equal 282
3.5 Chapter Review 283

CHAPTER 4 PLACE VALUE

4.1 Count and Compare Numbers Within 120 285
4.2 Tens and Ones 286
4.3 Finding Sums and Differences 287
4.4 Chapter Review 288

TABLE OF CONTENTS

CHAPTER 5 MEASUREMENT

5.1 Compare and Measure Lengths	290
5.2 Clock and Time	291
5.3 Chapter Review	292

CHAPTER 6 REPRESENT AND INTERPRET DATA

6.1 Organize and Represent Data in a Chart	293
6.2 Organize and Represent Data in a Table	295
6.3 Interpret Data with Up to Four Categories	296
6.4 Chapter Review	298

CHAPTER 7 GEOMETRY

7.1 Geometric Properties - 2D and 3D Shapes	299
7.2 Identify and Draw 2D and 3D Shapes	299
7.3 Create Composite Figures in Two or Three Dimensions	300
7.4 Chapter Review	303

CHAPTER 8 PARTITION SHAPES

8.1 Divide Circles and Rectangles into Two or Four Equal Parts	305
8.2 Describe Shares as Halves, Fourths, and Quarters	306
8.3. Identify Equal and Unequal Shares	307
8.4 Chapter Review	309

COMPREHENSIVE ASSESSMENT I	310
COMPREHENSIVE ASSESSMENT II	312

ANSWERS AND EXPLANATIONS

1. ADDITION WITHIN 20

1.1 ADDITION BASICS

1. **Answer: 13 balls.**
Explanation: There are 8 balls on the left side and 5 balls on the right.
8 + 5 = 13.
8 balls + 5 balls = 13 balls.

2. **Answer: 10 flowers.**
Explanation: There are 6 flowers on the left side and 4 flowers on the right.
6 + 4 = 10.
6 flowers + 4 flowers = 10 flowers.

3. **Answer: A**
Explanation: There are 5 motorcycles on the left side and 3 motorcycles on the right side.
5 + 3 = 8. Therefore, 5 motorcycles + 3 motorcycles = 8 motorcycles.

4. **Answer: 11 dots.**
Explanation: 5 dots on left side dice and 6 dots on right dice.
5 + 6 = 11.
5 dots + 6 dots = 11 dots.

5. **Answer: A**
Explanation: 5 Cones on the first row and 3 Cones on the second row.
5 + 3 = 8.
5 Cones + 3 Cones = 8 Cones.

6. **Answer: D**
Explanation: 5 + 4 = 9.

7. **Answer: 13.**
Explanation: 3 strawberries + 4 pineapples. 3 + 4 = 7.

8. **Answer: 8 dots.**
Explanation: 6 dots on the first dice and 2 dots on the second dice. 6 + 2 = 8.
6 dots + 2 dots = 8 dots.

9. **Answer: 8 fishes.**
Explanation: 2 fishes on the left side and 6 fishes on the right side. 2 + 6 = 8
2 fishes + 6 fishes = 8 fishes.

10. **Answer: 13**
Explanation: 9 + 4 = 13.

11. **Answer: D**
Explanation: 10 + 2 = 12.

12. **Answer: 12**
Explanation: 9 insects on the left side and 3 insects on the right side. 9 + 3 = 12.
9 insects + 3 insects = 12 insects.

13. **Answer: A**
Explanation:

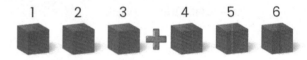

There are 6 Cubes.

14. **Answer: 17**
Explanation: 11 balls on the left side and 6 balls on the right side.
10 + 6 = 17.
11 Balls + 6 balls = 17 balls

15. **Answer: 18**
Explanation: 15 + 3 = 18.

16. **Answer: D**
Explanation: 9 balls in the first row and 6 balls in the second row.
9 + 6 = 15.
9 balls + 6 balls = 15 balls.

17. **Answer: 10**
Explanation: 7 + 3 = 10.

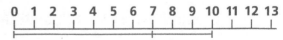

18. **Answer: B**
Explanation: 6 candies in the first row and 5 candies in the second row.
6 + 5 = 11. 6 candies + 5 candies = 11 candies.

ANSWERS AND EXPLANATIONS

19. Answer: C
Explanation: 7 on the left side and 6 flowers on the right side.
7 + 6 = 13 flowers.

20. Answer: A
Explanation: 9 + 7 = 16.

1.2 FINDING MISSING ADDEND

1. Answer: 5
Explanation: The number 5 makes the equations true.
5 + 5 = 10.

2. Answer: 9
Explanation: 9 apples and 3 bananas.
9 + 3 = 12, Missing number is 9.

3. Answer: 9
Explanation: 9 insects on the left side.
9 + 3 = 12, Missing number is 9.

4. Answer: A
Explanation: 7 + 7 = 14.

5. Answer: 6
Explanation: 6 chess pieces on the right side. 5 + 6 = 11, Missing number is 6.

6. Answer: 5
Explanation: 5 = 3 + 2, Missing number is 5.

7. Answer: B
Explanation: 9 + 2 = 11, Missing number is 2.

8. Answer: 2
Explanation: 7 + 2 = 9, missing number is 2.

9. Answer: B
Explanation: 5 + 6 = 11, Missing number is 5.

10. Answer: C
Explanation: 4 + 2 = 6, Missing number is 2.

11. Answer: D
Explanation: 4 + 4 = 8, Missing number is 4.

12. Answer: 3
Explanation: 5 + 3 = 8, 3 is the missing number.

13. Answer: A
Explanation: 6 + 2 = 8, Missing number is 2.

14. Answer: 3
Explanation: 3 flowers on the left side and 2 flowers on the right side.
3 + 2 = 5, 3 is the missing number.

15. Answer: B
Explanation: 7 + 3 = 10, Missing number is 7.

16. Answer: 7
Explanation: 7 flowers on the first row and 5 flowers on the second row. 16 = 7 + 5, 7 is the missing number.

17. Answer: A
Explanation: 11 + 7 = 18, can also be written as 18 − 7 = 11, 11 is the missing number.

18. Answer: A
Explanation: 1 + 6 = 7, 1 is the missing number.

19. Answer: C
Explanation: Total is 8, 3 flowers on the left side. 8 − 3 = 5,
it can also be written as 3 + 5 = 8, 5 is the missing number.

20. Answer: D
Explanation: 9 + 5 = 14, 5 is the missing number.

ANSWERS AND EXPLANATIONS

1.3 ADDITION WITH COUNTING ON AND MAKING TEN

1. Answer: B
Explanation: To find the total number of stickers Ben has using the counting-on method, we can add the number of orange stickers (8) and count on from there by adding the number of apple stickers (7). Starting with 8, we count on 7 more numbers: 9, 10, 11, 12, 13, 14, 15. Therefore, Ben has a total of 15 stickers (7 apple stickers + 8 orange stickers) using the counting-on method.

2. Answer: A
Explanation: To find the total number of fruits Alice has using the counting-on method, we can add the number of kiwis (6) and count on from there by adding the number of dragon fruits (3). Starting with 6, we count on 3 more numbers: 7, 8, 9. Therefore, Alice has a total of 9 fruits (6 kiwis + 3 dragon fruits) using the counting-on method.

3. Answer: 3
Explanation: Ava needs to make a group of 10. By recognizing that 7 is close to 10, Ava can count the missing marbles to find the answer 3.

4. Answer: 4
Explanation: Emily needs to make a group of 10. By recognizing that 6 is close to 10, Emily can count the missing candies to find the answer 4.

5. Answer: A
Explanation: To find the total number of fruits using the counting-on method, we can add the number of apples (6) and count on from there by adding the number of oranges (5). Starting with 6, we count on 5 more numbers: 7, 8, 9, 10, 11. Therefore, Tom has a total of 11 fruits (6 apples + 5 oranges) using the counting-on method.

6. Answer: 2
Explanation: The expression 4+8=12
It can be written as 10+2=12
Therefore, 4+8=10+2

7. Answer: B
Explanation: To find the total number of toys using counting, we can add the number of balls (9) and count on from there by adding the number of bats (4). Starting with 9, we count on 4 more numbers: 10, 11, 12, 13. Therefore, David has a total of 13 toys (9 balls + 4 bats) using the counting method.

8. Answer: 6
Explanation: Bella needs to make a group of 10. By recognizing that 4 is close to 10, Bella can count the missing pens to find the answer 6.

9. Answer: A
Explanation: From the given expression, 9+2=11. It can be written as 10+1=11. Therefore, 9+2=10+1. So option A is correct.

10. Answer: B
Explanation: 9+7=16.
It can be written as 10+6 =16; 10+6=9+7
So, the number 6 will be in the blank.

11. Answer: 10+2=12 beads
Explanation: Green beads: 5; Yellow beads: 7
The expression is 5+7. 5+7=12
The number 12 can be written as 10+2
Total: 10+2=12 beads.

12. Answer: C
Explanation: From the given expression, 8 + 5 = 13; It can be written as 10+3=13. Therefore, 8+5=10 + 3. So option C is correct.

13. Answer: D
Explanation: 9+9=18;
It can be written as 10+8 =18; 10+8=9+9
So the number 8 will be in the blank.

ANSWERS AND EXPLANATIONS

14. Answer: D
Explanation: 9+5=14;
It can be written as 10+4 =14; 10+4=9+5; So the number 4 will be in the blank.

15. Answer: 2
Explanation: Jerry needs to make a group of 10. By recognizing that 8 is close to 10, Jerry can count the missing toy cars to find the answer 2.

16. Answer: A
Explanation: To find the total number of kites using the counting-on method, we can add the number of green kites (8) and count on from there by adding the number of red kites (3). Starting with 8, we count on 3 more numbers: 9, 10, 11. Therefore, Dhoni has a total of 11 kites (3 red kites + 8 green kites) using the counting-on method.

17. Answer: 10+2=12 stickers
Explanation: Butterfly stickers: 7
Lion stickers: 5; 7 + 5 = 12 stickers
The expression 7+5 represents the sum of three addends, and the result is 10+2=12.

18. Answer: 10+5=15
Explanation: To solve the number sentence 9+6, we add the three numbers together.
9+6 = 15.
The 10+ fact related to this addition is:
10 + 5 = 15.

19. Answer: 10+1=11
Explanation: To solve the number sentence 7 + 4, we add the three numbers together. 7 + 4 = 11. The 10+ fact related to this addition is: 10 + 1 = 11.

20. Answer: 5
Explanation: George needs to make a group of 10. By recognizing that 5 is close to 10, George can count the missing crayons to find the answer 5.

1.4 ADDITION WORD PROBLEMS

1. Answer: A
Explanation: parrots: 4, Lovebirds: 8,
4 + 8 = 12, Antony has 12 birds.

2. Answer: C
Explanation: Cars in parking: 6, Cars on the road: 5, 6 + 5 = 11, Jack saw 11 cars.

3. Answer: B
Explanation: Sandwich: 6 dollars
Soda: 8 dollars, 8 + 6 = 14
Mike spent 14 dollars in all.

4. Answer: D
Explanation: Lotus: 5, Lilies: 8, 5 + 8 = 13
Jenny has 13 flowers.

5. Answer: A
Explanation: Mangoes: 5, Papayas: 6,
Oranges: 3, 5 + 6 + 3 = 14
Fruits bought by mom: 14 fruits

6. Answer: B
Explanation: Cupcakes with Jessy: 8
Cupcakes from her friend: 5
8 + 5 = 13, Jessy has 13 cupcakes.

7. Answer: D
Explanation: Cows: 6, Goats: 6,
2 more cows added, 6 + 6 + 2 = 14
Mellisa has 14 animals on her farm.

8. Answer: A
Explanation: Jonny: 9 crayons,
Steffi: 6 crayons, 9 + 6 = 15,
Altogether they have 15 crayons.

9. Answer: B
Explanation: Green beads: 5, Blue beads: 7,
Black beads: 6, 5 + 7 + 6 = 18
Teena has 18 beads with her.

10. Answer: D
Explanation: 6 + 4 = 10, Hen laid 10 eggs

11. Answer: A
Explanation: Angel: 3 mangoes, Mary: 5 more mangoes than angel, 3 + 5 = 8, Mary have 8 mangoes.

12. Answer: A
Explanation: Practice in this month: 11 days
Practice in next month: 8 days, 11 + 8 = 19
Total: 19 days.

13. Answer: C
Explanation: Burgers: 5, Pizzas: 4, Soda: 3
5 + 4 + 3 = 12, Jammie ordered 12 items.

14. Answer: A
Explanation: Tom: 8 seashells, Jerry: 2 seashells, 8 + 2 = 10, Sum: 10 seashells.

15. Answer: D
Explanation: William: 5, Chris: 2, 5 + 2 = 7
Total: 7 video games.

16. Answer: A
Explanation: Candles with Max: 8, Candles from her mom: 5, 8 + 5 = 13,
Max has 13 candles.

17. Answer: B
Explanation: Balls with Jack: 12, Balls from his friend: 7, + 7 = 19 , Jack has 19 balls.

18. Answer: B
Explanation: Red color eggs: 7, Yellow color eggs: 3, Blue color eggs: 2,
7 + 3 + 2 = 12, Total: 12 eggs.

19. Answer: C
Explanation: Balloons blew by Shyam: 13
Balloons blew by Steffi: 3,
Total: 13 + 3 = 16 Balloons.

20. Answer: D
Explanation: Fish in tank: 8
Fish added after some time: 5
4 more fish were added. 8 + 5 + 4 = 17.
There are 17 fish in the tank.

1.5 CHAPTER REVIEW

1. Answer: 14 butterflies
Explanation: There are 9 butterflies on the left side and 5 butterflies on the right. 9+ 5 = 14.
9 butterflies + 5 butterflies = 14 butterflies.

2. Answer: A
Explanation: Guests in the first hour: 5,
Guests in the second hour: 10, 5 + 10 = 15.
There are 15 guests.

3. Answer: B
Explanation: Airplanes in the airport: 9,
4 more Airplanes land in the same airport
9 + 4 = 13, 13 Airplanes are there in the airport.

4. Answer: B
Explanation: 8 rockets on the left side and 4 rockets on the right side. 8 + 4 = 12,
So option B is correct.

5. Answer: C
Explanation: Oranges from the first tree: 2
Oranges from the next tree: 5, Oranges from the last tree: 4, 2+5+4; The expression 2+5+4 represents the sum of three addends, and the result is 11; 2+5+4 = 11, Jessy picked 11 oranges altogether.

6. Answer: D
Explanation: Blue bikes: 8, Red bikes: 3, Green bikes: 3, 8 + 3 + 3, The expression 8+3+3 represents the sum of three addends, and the result is 14. 8 + 3 + 3 = 14, 10 + 4 = 14.

7. Answer: 8 + 5 = 13
Explanation: Sharpners in the first row: 8
Sharpners in the second row: 5
8 + 5 = 13, It can also be written as 10 + 3 = 13
There are 13 sharpners.

8. Answer: 10 + 5 = 15.
Explanation: 7+8 = 15. The number 15 can be written as 10+5. Therefore, 7+8 = 15, 10+5 = 15.

9. Answer: 12
Explanation: 7 + 5 = 12.

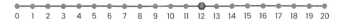

10. Answer: C
Explanation: Cars: 9; Bikes: 6; Buses: 4; 9 + 6 + 4. The expression 9 + 6 + 4 represents the sum of three addends, and the result is 19. So the option C is correct.

11. Answer: B
Explanation: Books on the left side: 6 Books on the right side: 3, The expression is 6 + 3, So the option B is correct. 6 + 3 = 9, There are 9 books.

12. Answer: C
Explanation: 5 + 4 + 9 = 18, The number 18 can be written as 10 + 8. Therefore, 10 + 8 = 5 + 4 + 9, 8 is the missing number.

13. Answer: 10
Explanation: 6 candies on the left side and 4 candies on the right side. 6 + 4 =10, There are 10 candies. 6 candies + 4 candies = 10 candies.

14. Answer: 7
Total: 12, Number on the left side: 5, Missing number: Total – Number on the left side, 12 – 5 = 7, 7 is the missing number. 5 + 7 = 12, So the answer 7 is correct.

15. Answer: 10 + 1 = 11.
Explanation: Balls in the first row: 8, Balls in the second row: 3, 8 + 3 = 11.
The number 11 can be written as 10 + 1. 10 + 1 = 11, There are 11 balls altogether.

16. Answer: 16
Explanation: 9 + 7 = 16.

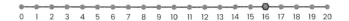

17. Answer: 15
Explanation: 9 on the left side and 6 crayons on the right side. 9 + 6 = 15, 15 is the missing number.

18. Answer: 10 + 2 = 12
Explanation: 9+3 = 12, The number 12 can be written as 10+2. Therefore, 9+3 = 12, 10+2 =12.

19. Answer: 10 + 3 = 13
Explanation: Initially, Angel has 7 teeth, and it is expected that 6 more teeth will grow after two years. Thus, the total number of teeth after two years can be determined by adding 7 and 6, resulting in 13. This can also be expressed as 10 + 3. Therefore, there will be a total of 13 teeth after two years.

20. Answer: 12
Explanation: Flowers in the first row: 7, Flowers in the second row: 5, 7 + 5 = 12, There are 12 flowers in total. 7 flowers + 5 flowers = 12 flowers.

ANSWERS AND EXPLANATIONS

2. SUBTRACTION WITHIN 20

2.1 SUBTRACTION BASICS

1. Answer: A
Explanation: Subtraction is nothing but "Taking away" or "how many are left" or "how many more or less." Thus, taking away 5 crayons from 18 crayons will give 13 crayons. That is, 18 - 5 = 13. The correct answer is A.

2. Answer: D
Explanation: Jessy has drawn 18 cakes. She now needs to cross out 6 cakes to represent the 6 cakes that were eaten. 18 – 6 = 12
12 cakes were left.

3. Answer: C
Explanation: Subtraction is nothing but "Taking away" or "how many are left" or "how many more or less." 15 – 5 = 10

4. Answer: C
Explanation: Subtraction is nothing but "Taking away" or "how many are left" or "how many more or less." Thus, taking away 3 apples from 12 apples will gives 9 apples. That is, 12-3=9.

5. Answer: 8
Explanation: Square blocks: 12,
Circle Blocks: 4
12 – 4 = 8, Thus, taking away 4 blocks from 12 blocks will give 8 blocks.

6. Answer: C
Explanation: Taking away 5 from 9 will give 4. 9 – 5 = 4, The missing number is 4.

7. Answer: D
Explanation: Roses with Robin: 9, He gave 3 roses to his mom. 9 – 3 = 6, Taking away 3 roses from 9 roses will give 6 roses.

8. Answer: D
Explanation: There are 12 hearts total with 3 of them separate. You can represent this as 12 -3. 12 – 3 = 9.

9. Answer: B
Explanation: The addition equation 6 + ? = 10 can be written as the subtraction equation 10 - 6 = ?.

10. Answer: 7
Explanation: There are 10 ice creams total with 3 of them separate. You can represent this as 10 – 3 = 7.

11. Answer: B
Explanation: Fishes with Jerry: 10, He gave 4 fishes to his friend. 10 – 4 = 6, taking away 4 fishes from 10 fishes will give 6 fishes.
6 fishes will remain in the aquarium.

12. Answer: A
Explanation: Lollipops with Rachel: 15
She ate 4 lollipops.
15 – 4 = 11, taking away 4 from 15 gives 11.
11 lollipops are there with Rachel.

13. Answer: B
Explanation: Peter caught 8 butterflies. He let 3 of them fly. 8 – 3 = 5, Taking away 3 from 8 is 5. Peter has 5 butterflies.

14. Answer: D
Explanation: Drawings with Ryan: 12, He gave some to his brother. Remaining drawings with Ryan: 5, 12 – 5 = 7, The missing number is 7.

15. Answer: A
Explanation: Total Teddy bears: 10
10 - 7 = 3, Take away 7 from 10 gives 3.
The missing number is 3.

16. Answer: 9
Explanation: 13 – 4 = 9, Taking away 4 blocks from 13 blocks will give 9 blocks.

ANSWERS AND EXPLANATIONS

17. Answer: D
Explanation: Barbie toys with Jessy: 8, Barbie toys from her friend: 2, 8 − 2 = 6, Taking away 2 toys from 8 toys will give 6 toys.

18. Answer: B
Explanation: 10 − 7 = 3
Taking away 7 from 10 will give 3.
7 + 3 = 10 can also be written as 10 − 7 = 3

19. Answer: D
Explanation: Total books: 9, 3 of them separate. You can represent this as 9 - 3.
9 − 3 = 6, Taking away 3 from 9 to give 6.

20. Answer: A
Explanation: 10 - 4 = 6, Take away 6 from 10 gives 4. The missing number is 4.

2.2 SUBTRACTION BY DECOMPOSING A NUMBER INTO TEN

1. Answer: A
Explanation: It is given that Simon bought some marbles. He gave $ 18 and the shopkeeper gave him $ 11. In order to find the cost of the marbles, subtract 11 from 18 to get 7. The cost of the marbles is $ 7.

2. Answer: 4
Explanation: It is given that Emily purchased some flowers. She gave $ 14 and the shopkeeper gave him $ 10.
subtract 10 from 14 to get 4.
The cost of the flowers is 4.

3. Answer: C
Explanation: Mangoes in a basket = 15
Rotten mangoes = 7
Good mangoes = subtract 7 from 15 to get 8.
Therefore, there are 8 good mangoes.

4. Answer: D
Explanation: David bought a gift box: $ 10
Amount paid by David: $ 19
Change received by David:
Subtract 10 from 19 to get 9.
David gets $9.

5. Answer: C
Explanation:
Sam bought a chocolate box: $ 10
Amount paid by Sam: $ 11
Change received by David: Subtract 10 from 11 to get 1. Sam gets $1.

6. Answer: 9
Explanation: Apples in a basket = 18
Rotten apples = 9
Good apples = subtract 9 from 18 to get 9.
Therefore, there are 9 good apples.

7. Answer: $ 3
Explanation: It is given that Tom bought some bananas. He gave $ 15 and the shopkeeper gave him $ 12. In order to find the cost of the bananas, subtract 12 from 15 to get 3. The cost of the bananas is $ 3.

8. Answer: 9
Explanation: Cherries in a basket = 16
Rotten cherries = 7; Good cherries = subtract 7 from 16 to get 9. 10 − 1 = 9.
Therefore, there are 9 good cherries.

9. Answer: $ 4
Explanation: It is given that Jerry bought some roses. He gave $ 11 and the shopkeeper gave him $ 7. In order to find the cost of the roses, subtract 7 from 11 to get 4.
The cost of the roses is $ 4.

10. Answer: $5
Explanation: Peter bought a toy car: $ 12
Amount paid by Peter: $ 17 ;
Change received by Peter: Subtract 12 from 17 to get 5. Peter gets $5.

ANSWERS AND EXPLANATIONS

11. Answer: 8
Explanation: Total wooden rods: 12
Broken wooden rods : 4
Not wooden rods: Subtract 4 from 12 to get 8.
There are 8 not broken wooden rods.

12. Answer: D
Explanation: Total cookies with Jack: 15
Broken cookies: 6 ; Not broken cookies:
Subtract 6 from 15 to get 9.
There are 9 cookies that are not broken.

13. Answer: A
Explanation: Total flowers in a plant: 15
Flowers picked by Sam: 11
Flowers left on the plant:
Subtract 11 from 15 to get 4.
There are 4 flowers left on the plant.

14. Answer: 9
Explanation: Total balloons: 14
Broken balloons : 5;
Not broken balloons:
Subtract 5 from 14 to get 9.
9 balloons are not broken.

15. Answer: 6
Explanation: Subtract 9 from 15 to get 6.

16. Answer: 3
Explanation:
Subtract 7 from 10 to get 3, 10 − 7 = 3

17. Answer: 8
Explanation: Total number of ducks: 11,
Subtract 3 from 11 to get 8.
Therefore, 11 − 3 = 8

18. Answer: D
Explanation: Points scored by Tom: 14
Points scored by James: 5 less than Tom
Subtract 5 from 14 to get 9.

19. Answer: D
Explanation: Total Kiwis with Max: 11
Kiwis ate by Max: 5
Remaining kiwis with Max: 11 - 5 = 6.

ANSWERS AND EXPLANATIONS

20. Answer: D
Explanation: Oranges to his neighbor's: 3
Oranges to his mother: Subtract 3 from 9 to get 6. 9 − 3 = 6.

2.3 SUBTRACTION WORD PROBLEMS

1. Answer: A
Explanation: Reena: 8 coins, Robin: 2 coins
8 − 2 = 6 coins, Taking away 2 from 8 gives 6.
Reena has 6 more coins than Robin.

2. Answer: B
Explanation: Car toys with Joey: 10, He lost 3 of them. 10 − 3 = 7, Taking away 3 from 10 gives 7. Joey has 7 toys left with him.

3. Answer: C
Explanation: Gift boxes with Andria: 17
She distributed 9 gift boxes 17 − 9 = 8, Taking away 9 from 17 gives 8. 8 gift boxes are left.

4. Answer: A
Explanation: Total beads: 10,
Green beads: 2, Blue beads: 4
10 − 4 − 2 = 4, Take away 6 from 10 to give 4.
10 beads − 4 beads − 2 beads = 4 red beads

5. Answer: D
Explanation: Cookies with Maria: 13
She gave 3 cookies to his friends.
13 − 3 = 10, Take away 3 from 13 gives 10.
Maria has 10 cookies.

6. Answer: 8
Explanation: Christina bought 11 stars. She kept 3 stars with her. 11 − 3 = 8, Take away 3 from 11 to get 8. She gave 8 stars to her friend.

7. Answer: C
Explanation: 18 − 6 = 12
Angel needs $ 12 more dollars.

ANSWERS AND EXPLANATIONS

8. Answer: A
Explanation: Tom has 6 car toys. His friend came cover with some more car toys. Together, they have 10 car toys. Take away 6 car toys from 10 car toys to get the number of car toys brought by his friend. 10 − 6 = 4
4 car toys are brought by his friend.

9. Answer: D
Explanation: Strawberries with Mouna: 10
She ate 3 in the morning. 10 − 3 = 7, Take away 3 from 10 gives 7.
Mouna ate 7 strawberries in the evening.

10. Answer: A
Explanation: Total: 12 people; Boys: 8;
12 − 8 = 4, Taking away 8 from 12 gives 4.
There are 4 girls.

11. Answer: 3 apples
Explanation: Charles bought 4 apples, his mom gave him some more. Now he has 7 apples. Number of apples given by Charles's mom = 7 − 4 = 3.

12. Answer: D
Explanation: Jerry has drawn 12 burgers. He now needs to cross out 5 burgers to represent the 5 burgers that were eaten.
12 − 5 = 7, Take away 5 from 12 give 7.
7 burgers were left.

13. Answer: B
Explanation: Sam: 10 video games
Sarah: 4 video games, 10 − 4 = 6
Take away 4 from 10 gives 6.
Sarah has 6 video games more than Sam.

14. Answer: C
Explanation: Lovebirds: 11, Parrots: 5
11 − 5 = 6, Take away 5 from 11 gives 6.
Tim has 6 lovebirds more than parrots.

15. Answer: D
Explanation: Total students: 25; Boys: 15
Girls: 10; 15 − 10 = 5, Take away 10 from 15 gives 5. There are 5 more boys than girls.

16. Answer: A
Explanation: Mary: $ 8
Mary paid 13 − 8 = 5
Mary paid $ 5 more than Steffi.

17. Answer: C
Explanation: Pinky: 12 marks, and Tim: 19 marks, Pinky scored the least marks: 12
19 − 12 = 7

18. Answer: A
Explanation: Sam: 7; Linton: 3; 7 − 3 = 4,
Take away 3 from 7 gives 4.
Sam has 4 mangoes more than Sam

19. Answer: A
Explanation: Simon had 15 pancakes. He gave some to his brother.
Now he has 7; 15 − 7 = 8
Simon gave 8 pancakes to his brother.

20. Answer: D
Explanation: Blessy planted: 19 trees
Angel planted: 10 trees. 19 − 10 = 9
Blessy planted 9 more trees than Angel.

2.4 CHAPTER REVIEW

1. Answer: A
Explanation: Erasers with Brito: 13. He gave 9 erasers to his friend. 13 − 9 = 4, Take away 9 from 13 gives 4. erasers are left with Brito.

2. Answer: 3
Explanation: 9 cupcakes on the left side and 6 cupcakes on the right side.
9 − 6 = 3, Take away 6 from 9 gives 3.
9 cupcakes − 6 cupcakes = 3 cupcakes.

3. Answer: B
Explanation: Red balls: 14, Yellow balls: 9,
14 − 9 = 5, Take away 9 from 14 gives 5.
There are 5 more red balls than yellow balls.

4. Answer: A
Explanation: 15 − 6 = 9, Take away 6 from 15 gives 9. The missing number is 9.

5. Answer: D
Explanation: Flowers in the first: 8, Flowers in the second: 5, Total flowers: 13, Possible expressions: 13 – 8 = ? or 13 – 5 = ?, So D is the correct answer. 13 – 5 = 8

6. Answer: C
Explanation: Tickets sold by Simon: 9, Tickets sold by Pinky: 3, 9 – 3 = 6, Take away 3 from 9 gives 6. Pinky sold 6 tickets less than Simon.

7. Answer: 5
Explanation: 11 – 6 = 5, Take away 6 from 11 gives 5. 5 is the missing number.

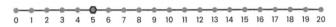

8. Answer: 7
Explanation: Car toys in the first row: 7, Car toys in the second row: 2, Total car toys: 9, 9 – 2 = 7, the missing number is 7.

9. Answer: D
Explanation: Option A: 10 – 4 = 6 – True
Option B: 12 – 5 = 7 – True
Option C: 4 + 10 = 14 - True
Option D: 11 - 4 = 7
the equation given is 11 - 4 = 14,
Option D is false.

10. Answer: 7
Explanation:
Subtract 3 from 10 to get 7, 10 - 3 =7.

11. Answer: D
Explanation: Fluffy caught: 10 butterflies
Puffy caught: 17 butterflies.
17 – 10 = 7, Take away 10 from 17 gives 7.
Fluffy caught 7 butterflies, fewer than the Puffy.

12. Answer: 13
Explanation: 18 – 5 = 13, Take away 5 from 18 gives 13. 13 is the missing number.

13. Answer: 4
Explanation: Total Ice creams: 10
Take away: 4, 10 – 4 = 6, take away 4 from 10 gives 6. Related sentence is 10 – 6 = 4
6 is the missing number.

14. Answer: A
Explanation: Steps by Edwin: 13, Steps by Allen: 5, 13 – 5 = 8,
Take away 5 from 13 gives 8. 8 steps more.

15. Answer: 12 – 4 = 8 = 10 - 2
Explanation: Total balloons : 12
Take away 4 from 12 gives 8, Decompose the number 8 leading to ten as 10 - 2.
Therefore, 12 – 4 = 8 = 10 – 2

16. Answer: D
Explanation: 18 – 7 = 11, Subtract 7 from 18 gives 11. Related sentence is 18 – 11 = 7
11 is the missing number.

17. Answer: C
Explanation: Jonny needs to put 19 color pencils on the table. He put 6 color pencils on the table already. 19 – 6 = 13, Take away 6 from 19 gives 13. Jonny needs to put 13 more color pencils on the table.

18. Answer: 9
Explanation: Cotton candies with Laura: 14
Cotton candies with Tim: 5,
14 – 5 = 9,
Subtract 5 from 14 gives 9.
Tim has 9 less cotton candies than Laura.

19. Answer: 7
Explanation: 14 – 6 = 8,
take away 6 from 14 gives 8.
8 is the missing number.

20. Answer: D
Explanation: Strawberries bought by Tom: 9
Strawberries given to his neighbors: 3
9 – 3 = 6, Take away 3 from 9 gives 6.

ANSWERS AND EXPLANATIONS

3. PROPERTIES OF ADDITIVE AND SUBTRACTIVE OPERATIONS AND EQUATIONS

3.1 COMMUTATIVE PROPERTY AND ASSOCIATIVE PROPERTY OF ADDITION

1. **Answer: 8 fruits.**
Explanation: There are 4 mangoes + 4 strawberries = 8 fruits.
It can be expressed as 4 strawberries + 4 mangoes = 8 fruits.
Both sides are equal.
So the answer is 4 mangoes + 4 strawberries = 8 fruits = 4 strawberries + 4 mangoes.

2. **Answer: B**
Explanation: White shirts: 5
Black shirts: 8
5 white shirts + 8 black shirts = 13 shirts.
Peter bought 13 shirts in total.

3. **Answer: 7 insects.**
Explanation:
There are 4 butterflies + 3 bees = 7 insects. It can be expressed as 3 bees + 4 butterflies = 7 insects.
Both the sides are equal. So the answer is 4 butterflies + 3 bees = 7 insects = 3 bees + 4 butterflies.

4. **Answer: D**
Explanation: Given expression: 9 + 7 = 16
It only matches with the expression 16 = 7 + 9
Thus the addition sentence 9+7=16 matches with 16 = 7 + 9.

5. **Answer: 14**
Explanation: There are 8 gold fishes + 6 silver fishes = 14 fishes, It can also be expressed as 6 silver fishes + 8 gold fishes = 14 fishes. So, the answer is 8 + 6 =14; 6 + 8 = 14. Jessy has 14 fishes in total.

6. **Answer: C**
Explanation: Eggs in the refrigerator: 9, Eggs outside: 6, 9 eggs + 6 eggs = 15 eggs.
It can also expressed as 10 + 5 = 15.
Claire has 15 eggs.

7. **Answer: A.**
Explanation: Sunflowers: 8; Roses: 7;
8 sunflowers + 6 roses = 14 flowers
Tim buys 14 flowers in total.

8. **Answer: A**
Explanation: 3+6=9
Given expression: 9 = 3 + 6
It can also be written as 6 + 3 = 9.
Thus the addition sentence 9 = 3 + 6 matches with 6 + 3 = 9.

9. **Answer: Many answers.**
Explanation: There are 3 lollipops and 5 candies = 8, In addition sentence it can also be written as
3 + 5 = 8 | 5 + 3 = 8 | 8 = 3 + 5 | 8 = 5 + 3.

10. **Answer: 12**
Explanation: 7 + 5 = 12

12 is the answer.

11. **Answer: A**
Explanation: Red roses: 7
Pink roses: 11.
7 red roses + 11 pink roses = 18 roses.
It can also be expressed as 10 + 8 = 18.
Tom bought 18 roses in total.
So, option A is correct.

12. **Answer: A**
Explanation: There are 3 dolls + 2 teddy bears + 2 toy cars = 7 toys.
3 + 2 + 2 = 7, Option A is correct.

13. **Answer: 19 pets**
Explanation: Blessy has 7 sparrows + 5 parrots + 7 lovebirds = 19 pets
7 + 5 + 7 = 19
Blessy has 19 pets.

ANSWERS AND EXPLANATIONS

14. Answer: B
Explanation: They worked 8 hours + 5 hours + 3 hours = 16 hours.
8 + 5 + 3 = 16
16 hours.

15. Answer: 12
Explanation: There are 5 red apples + 7 green apples = 12 apples; It can also be expressed as 7 green apples + 5 red apples = 12 apples.
So, the answer is 5 + 7 = 12; 7 + 5 = 12.
John has 12 apples in total

16. Answer: 16 pens
Explanation: Maria has 8 blue pens + 5 black pens + 3 red pens = 16 pens
8+5+3=16
So, the answer is 8+5+3=10+6=16.
Maria has 16 pens.

17. Answer: 10 fruits
Explanation: There are 6 apples + 3 oranges + 1 watermelon = 10 fruits.
6 + 3 + 1 = 10.
6 + 1 + 3 = 10

18. Answer: C
Explanation: Garden has 7 red vinca plants + 3 pink vinca plants + 8 white vinca plants = 18 plants.
7+3+8=18
There are 18 vinca plants in the garden.

19. Answer: 17
Explanation: Tomatoes : 9, Cabbages: 8
9 tomatoes + 8 cabbages = 17 vegetables
Emy purchased 17 vegetables in total.
9 + 8 =17
8 + 9 = 17.

20. Answer: 14 balloons.
Explanation: There are 6 round balloons + 5 heart balloons + 3 star balloons.
6 + 5 + 3 =14, So the answer is 6 round balloons + 5 heart balloons + 3 star balloons = 14 balloons.

3.2 THREE NUMBER ADDITION OR SUBTRACTION

1. Answer: 13 vegetables
Explanation: There are 4 Broccoli, 5 red Capsicums and 4 Beetroots. That is 4+5+4=13
Thus, 4 Broccoli + 5 red Capsicums + 4 Beetroots = 13 Vegetables.

2. Answer: A
Explanation: Eggs cooked in morning: 4
Eggs cooked in evening: 8
Eggs cooked in night: 3
4 + 8 + 3 = 15; Maria cooked 15 eggs.

3. Answer: 13 plants
Explanation: Rose plants: 9, Jasmine plants: 6
He picked out 2 plants. 9 + 6 - 2 = 13.
There are 13 plants on the front yard.

4. Answer: B
Explanation: Cookies baked in morning: 5
Cookies baked in evening: 9 ;
She ate 3 cookies at night. 5 + 9 - 3 = 11.
There are 11 cookies with Tina.

5. Answer: 14 fishes
Explanation: Red fish: 8 ; Black fish: 7
He removed 1 fish. 8 + 7 - 1 = 14.
Jack has 14 fish.

6. Answer: 5 vegetables
Explanation: There are 4 cabbages, 3 tomatoes and 4 carrots. That is 4 – 3 + 4 = 5
Thus, 4 cabbages - 3 tomatoes + 4 carrots = 5 Vegetables.

7. Answer: A
Explanation: There are 5 pink lotus, 2 white lotus and 4 lillises. 5 + 2 + 4 = 11, It can also be written as 10 + 1 = 11.
There are 11 flowers in the pond

8. Answer: C
Explanation: Blue pens: 8; Black pens: 5;
He missed 3 pens. 8+5−3=10, Ben has 10 pens.

ANSWERS AND EXPLANATIONS

9. Answer: B
Explanation: Here, 8 + 3 - 4 = 7. The number 7 can be written as 10 - 3 = 7. The completed number sentences 8 + 3 - 4 = 7, 10 - 3 = 7.

10. Answer: C
Explanation: There are 5 Candies, 8 biscuits and 5 chocolates. 5 + 8 + 5 = 18. Mellisa has 18 items with her.

11. Answer: A
Explanation: Here, 7 + 4 - 2 = 9. The number 9 can be written as 10 - 1 = 9. The completed number sentences 7 + 4 - 2 = 9, 10 - 1 = 9.

12. Answer: C
Explanation: Here, 3 + 2 + 5 = 10 and 10 + 0 = 10. Thus the sentence 3 + 2 + 5 = 10 matches only with 10 + 0 = 10.

13. Answer: 12 fruits
Explanation: There are 5 green apples, 3 oranges and 4 strawberries. That is 5 + 3 + 4 = 12, Thus, 5 green apples + 3 oranges + 4 strawberries = 12 fruits.

14. Answer: B
Explanation: There are 9 parrots and 3 pigeons. Four of them flew away. 9 + 3 − 4 = 8 It can also be written as 10 − 2 = 8. That is 9 + 3 − 4 = 10 − 2 = 8. There are 8 birds on the tree.

15. Answer: D
Explanation: Lions: 8, Tigers: 4, Deers: 5 That is 8 + 4 + 5 = 17. Therefore, Jeo saw 17 animals.

16. Answer: A
Explanation: Green pencils: 7 Blue pencils: 9; He misplaced 4 pencils. 7 + 9 − 4 = 12 ; Alan has 12 pencils.

17. Answer: D
Explanation: Here, 5 + 2 + 5 = 12 and 12 = 10 + 2. Thus the sentence 5 + 2 + 5 = 12 matches only with 12 = 10 + 2.

18. Answer: 17 marbles.
Explanation: Ronald bought 7 pink marbles, 5 green marbles and 5 yellow marbles. 7 + 5 + 5 = 17. Ronald bought 17 marbles altogether

19. Answer: B
Explanation: Joseph made 8 red kites and 4 yellow kites. He gave 3 to his friend. 8 + 4 − 3 = 9 , It can also be written as 10 − 1 = 9, That is 8 + 4 − 3 = 10 − 1 = 9 9 kites are are with Joseph.

20. Answer: 14
Explanation: There are 6 shirts, 5 gowns and 3 pants. 6 + 5 + 3 = 14. Mercy has 14 items with her.

3.3 ADD OR SUBTRACT WITHIN 100 USING PROPERTIES OF OPERATIONS

1. Answer: A
Explanations: 52 children were standing in line, and 25 children joined them. 52 + 25 = 77; 77 Children are in line.

2. Answer: 69
Explanations: 48 mangoes in the basket and 21 mangoes added in the same basket. 48 − 21 = 69; 69 mangoes in the basket.

3. Answer: B
Explanations: 10 + 8 = 18 Stella have 18 cupcakes.

4. Answer: B
Explanations: 70 − 56 = 14 So, the answer is 14.

5. Answer: D
Explanation: 2 stick of 10 cubes = 20 ones His mom gave him 9 ones. 20 + 9 = 29. Mark has 29 cubes.

ANSWERS AND EXPLANATIONS

6. Answer: 2 tens.
Explanations: Totally there are 28 stars.
28 is 2 tens and 8 ones.
So, Jenny can make 2 tens.

7. Answer: Missing numbers are 35 and 72.
Explanations: Here 45−10=35 and 62+10=72
So, the missing numbers are 35 and 72.

8. Answer: A
Explanations: April has 4 piles of 10 and 3 loose stamps: 40 + 3 =43 stamps.
She gave 1 pile of 10 and 3 loose stamps to her friend: 10 + 3 = 13 stamps.
So, 43 − 13 = 30 stamps.
April has 30 stamps with her.

9. Answer: Missing numbers are 14 and 44.
Explanations: Here 24−10=14 and 34+10=44.
So, the missing numbers are 14 and 44.

10. Answer: B
Explanations: 9 tens and 2 ones = 90+2=92.
So, the statement is false.

11. Answer: 66
Explanations: 4 tens is 40 and 6 ones is 6.
40 + 6 = 46; Two tens = 20 ; 46 + 20 = 66.
The number is 66.

12. Answer: A
Explanations: 38 carrots and 26 cabbages = 38+26=64; Sharon has 64 vegetables.

13. Answer: 37 pages
Explanations: 17+20=37 pages.
Sam write 37 pages.

14. Answer: 34
Explanations: 6 tens and 10 ones =60+10=70.
3 tens and 6 ones=30+6=36.
70−36=34. 34 means 3 tens and 4 ones.

15. Answer: C
Explanation: 53 − 31 = 22
So, the answer is 22.

16. Answer: 35 balls.
Explanation: Ms. Edwin had 45 balls.
10 have been lost. 45 − 10 = 35.
There are 35 balls.

17. Answer: 24
Explanation: 4 tens and 8 ones = 40+8= 48
2 tens and 4 ones = 20+4=24
48 − 24 = 24
24 means 2 tens and 4 ones.

18. Answer: A
Explanation: 5 tens is 50 and 1 ones is 1.
50 + 1 = 51
Two tens = 20
51 + 20 = 71
The number is 71.

19. Answer: B
Explanation: 4 tens and 12 ones = 40 + 12 = 52. So, the statement is false.

20. Answer: 21
Explanation: 3 tens and 9 ones = 30 + 9 = 39
1 tens and 8 ones = 10 + 8 = 18
39 − 18 = 21; 21 means 2 tens and 1 ones.

3.4 FINDING TWO VALUES ARE EQUAL

1. Answer: 4 + 4 = 5 + 3.
Explanation: There are 4 oranges + 4 strawberries = 8 fruits on the left side.
There are 5 oranges + 3 strawberries = 8 fruits on the right side. Both sides are equal.
So, the answer is 4 + 4 = 5 + 3.

2. Answer: A
Explanation: Here 7 + 2 = 9, It also matches with 5 + 4 = 9, Both values are the same.
So, option A is true.

3. Answer: Mike is right.
Explanation: First expression: 15 − 2 = 13
Second expression: 19 − 6 = 13
Both are equal, So, Mike is right.

ANSWERS AND EXPLANATIONS

4. Answer: A
Explanation: Here, 9 + 2 = 11
It only matches with 3 + 8 = 11
8 is the missing number.

5. Answer: A
Explanation: Here, 10 + 3 = 13 and 15 − 2 = 13 are equal. So, the equation is true.

6. Answer: C
Explanation: Here, 12 − 5 = 7, Therefore, the expression 12 − 5 matches with 7.

7. Answer: 7 + 2 = 4 + 5.
Explanation: There are 7 sunflowers + 2 roses = 9 flowers on the left side.
There are 4 sunflowers + 5 roses = 9 flowers on the right side. Both sides are equal.
So, the answer is 7 + 2 = 4 + 5.

8. Answer: B
Explanation: Here 9 + 3 = 12, It also matches with 6 + 6 = 12, Both values are the same. So, option B is true.

9. Answer: Jessy is right.
Explanation: First expression: 11 − 2 = 9
Second expression: 18 − 9 = 9
Both are equal, So, Jessy is right.

10. Answer: B
Explanation: Here, 6 + 3 = 9 is not equal to 19 − 8 = 11. So, the equation is false.

11. Answer: A
Explanation: Here, 2 + 5 = 7 is equal to 3 + 4 = 7. So, the equation is True.

12. Answer: D
Explanation: Here, 3 + 4 = 7, It only matches with 5 + 2 = 7, So, 5 is the missing number.

13. Answer: 5 + 1 = 3 + 3.
Explanation: There are 5 butterflies + 1 bee = 6 insects on the left side. There are 3 butterflies + 3 bees = 6 insects on the right side. Both sides are equal.
So, the answer is 5 + 1 = 3 + 3.

14. Answer: C
Explanation: Here 3 + 2 = 5,
It also matches with 4 + 1 = 5, Both values are the same. So, option C is true.

15. Answer: Sandra is right.
Explanation: First expression: 17 − 5 = 12
Second expression: 18 − 6 = 12,
Both are equal, So, Sandra is right.

16. Answer: A
Explanation: Here, 3 + 3 = 6 and 11 − 5 = 6 are equal. So, the equation is true.

17. Answer: B
Explanation: Here, 2 + 4 = 6, It only matches with 3 + 3 = 6, So, 3 is the missing number.

18. Answer: C
Explanation: Here, 4 + 8 = 12
It only matches with 5 + 7 = 12
So, 5 is the missing number.

19. Answer: A
Explanation: Here, 9+2=11 and 13−2=11
So, the only expression 9+2 is equal to 13−2.

20. Answer: B
Explanation: Here, 4 + 3 = 7 and 12 − 6 = 6 are not equal.
So, the equation is false.

3.5 CHAPTER REVIEW

1. Answer: 7 vegetables.
Explanation: There are 4 red capsicums + 3 onions = 7 vegetables.
It can be expressed as 3 onions + 4 red capsicums = 7 vegetables.
Both sides are equal.
So the answer is 4 red capsicums + 3 onions = 7 vegetables = 3 onions + 4 red capsicums.

ANSWERS AND EXPLANATIONS

2. Answer: A
Explanation: Given expression: 6 + 5 = 11
It only matches with the expression 11 = 7 + 4
Thus the addition sentence 6 + 5 = 11 matches with 11 = 7 + 4.

3. Answer: 7 + 4 + 7 = 18, 10 + 8 = 18
Explanation: There are 7 small crabs + 4 big crabs + 7 black crabs = 18 crabs
7 + 4 + 7 = 18,
It can also be written as 10 + 8 = 18.
So, the answer is 7 + 4 + 7 = 10 + 8 = 18.
There are 18 crabs on the beach.

4. Answer: B
Explanation: James bought 7 green apples and 5 mangoes. 7 green apples + 5 mangoes = 12 fruits
James buys 12 fruits in total.

5. Answer: 17
Explanation: 8 + 9 = 17

17 is the answer.

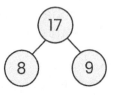

6. Answer: D
Explanation: Here, 5 + 3 + 9 = 17.
The number 17 can be written as 17 = 10 + 7.
Therefore, the additional sentence
5 + 3 + 9 = 17 matches with 17 = 10 + 7.
So, option D is correct.

7. Answer: A
Explanation: There are 5 red balloons + 3 green balloons + 4 yellow balloons = 12 balloons, 5 + 3 + 4 = 12, Option A is correct.

8. Answer: B
Explanation: Here 8 + 3 = 11, It only matches with 6 + 5 = 11, Both values are the same. So, option B is true.

9. Answer: Angel is right.
Explanation: First expression: 11 − 2 = 9
Second expression: 15 − 6 = 9
Both are equal, So, Angel is right.

10. Answer: C
Explanation: Here 12 + 3 = 15,
It only matches with 8 + 7= 15, Both values are the same. So, option C is true.

11. Answer: C
Explanation: Here 8 + 7 = 15,
It only matches with 12 + 3 = 15, Both values are the same. So, option C is true.

12. Answer: 4 + 3 = 2 + 5.
Explanation: There are 4 cabbages + 3 carrots = 7 vegetables on the left side.
There are 2 cabbages + 5 carrots = 7 vegetables on the right side.
Both sides are equal.
So, the answer is 4 + 3 = 2 + 5.

13. Answer: A
Explanation: Here, 12 + 5 = 17 and 19 − 2 = 17 are equal. So, the equation is true.

14. Answer: C
Explanation: There are 7 butterflies and 5 bees in the garden. Two of them flew away. 7 + 5 − 2 = 10. There are 10 insects in the garden.

15. Answer: B
Explanation: Here, 4 + 7 + 8 = 19 and 19 = 10 + 9. Thus the sentence 4 + 7 + 8 = 19. matches only with 19 = 10 + 9.

16. Answer: 11 dresses
Explanation: She gets 6 pink dresses and 8 blue dresses. She gave 3 of them to her sister. 6 + 8 − 3 = 11. Maria has 11 dresses with her now.

17. Answer: 8 insects.
Explanation: There are 7 butterflies, 3 bees, and 4 bugs. That is 7 − 3 + 4 = 8. Thus, 7 butterflies - 3 bees + 4 bugs = 8 insects.

ANSWERS AND EXPLANATIONS

18. Answer: A
Explanation: It is given that 8 + 2 + 3 = 13.
The number 13 can be written as 10 + 3 = 13.
So the answer is 3.

19. Answer: B
Explanation: The unknown-addend problem 11 + ? = 15 can be rewritten as the subtraction equation 15 - 11 = ?

20. Answer: D
Explanation:
Given the expression is 6 + 5 + 4 = ? to make the equation to 10+5=?, two terms need to be combined to make 10.
Add 6 and 4 to make 10.
Thus 6 + 5 + 4 = 10 + 5 = 15.

4. PLACE VALUE

4.1 COUNT AND COMPARE NUMBERS WITHIN 120

1. Answer: 35 flowers.
Explanation: There is 10 flowers in the first, second, and third row: a total of 30 flowers.
5 flowers in the fourth row.
Total: 30 + 5 = 35 flowers.

2. Answer: 119, 120.
Explanation: The number next to 118 is 119.
The number next to 119 is 120.
So the answer is 119, 120.

3. Answer: 55, 78, 82
Explanation: The smallest number is 55, the next smallest number is 78 and the greatest number is 82. So the answer is 55, 78, 82.

4. Answer: 43 butterflies.
Explanation: There is 5 rows, with 8 butterflies in each row: 40 butterflies. 3 butterflies in the fifth row. Total: 40 + 3 = 43 flowers.

5. Answer: C
Explanation: The number before 120 is 119.
119 is the answer. Thus, the option C Is correct.

6. Answer: 116, 119
Explanation: The number next to 115 is 116.
The number next to 118 is 119.
So the answer is 116, 119.

7. Answer: 17, 29, 49
Explanation: The smallest number is 17, the next smallest number is 29 and the greatest number is 49. So the answer is 17, 29, 49.

8. Answer: D
Explanation: There is 35 stars, add 15 stars to get 50 stars. 35 + 15 = 50 stars.
So, option D is correct.

9. Answer: A
Explanation: The number next to 115 is 116.
So she will count 116. Thus, option A is correct.

10. Answer: 55, 58.
Explanation: The number before 56 is 55.
The number next to 57 is 58.
So the answer is 55, 58.

11. Answer: 20 > 15
Explanation: There is 20 balls and 15 balls.
Comparing, 20 is greater than 15.
The answer is 20 > 15.

12. Answer: A
Explanation: There is 95 on the left side and 100 on the right side. Comparing 95 is less than 100. So, 95 is smaller than 100. 95 < 100, So option A is correct.

13. Answer: 41 < 52 < 63 or 63 > 52 > 41
Explanation: Here 41 is less than both 52 and 63, so 41 < 52 and 63. Then 52 is less than 63, so 52 < 63. Therefore 41 < 52 < 63 And also 63 is greater than 52 and 41, so 63 > 52 is greater than 41, 52 > 41. Therefore 63 > 52 > 41.

14. Answer: 8 < 10
Explanation: There is 8 red capsicums and 10 green capsicums. On comparing, 8 is smaller than 10. So, 8 is less than 10.
The answer is 8 < 10.

ANSWERS AND EXPLANATIONS

15. Answer: B
Explanation: There is 107 on the left side and 107 on the right side. Upon comparing 107 is equal to 107. So, 107 = 107.
Option B is correct.

16. Answer: B
Explanation: Pages read by Ronald: 75
Pages read by David: 89
Here, 89 is greater than 75, that is 89 > 75.
So, David read more pages.

17. Answer: 8 = 8
Explanation: There is 8 apples on the left and 8 oranges on the right sides 8 is equal to 8.
The answer is 8 = 8.

18. Answer: C
Explanation: There is 115 on the left side and 103 on the right side. Upon comparing 115 is greater than 103. So, 115 > 103.
Option C is correct.

19. Answer: A
Explanation: Here, the numbers greater than 45 are 55, 63, 82 and 71. But the number less than 60 is 55.
So, the answer is 55. Option A is correct.

20. Answer: C
Explanation: There is 15 on the left side and 10 on the right side.
Comparing 15 is greater than 10.
So, 15 > 10. Option C is correct.

4.2 TENS AND ONES

1. Answer: The place value of 7̲7 is tens.
Explanation: In a two-digit number the first position is tens digit where the tens number is written and the second position is ones digit where ones number is written.
The given number is 7̲7 in which the number 7̲ is underlined. Therefore, the place value of the given number 7̲ is tens.

2. Answer: 4 tens and 2 ones
Explanation: The given number is 42. It is known that, in a two-digit number the first position is called the ten's digit where the tens number is written and the second position is called the ones digit where ones number is written. That is, note that here the number in the ten's digit is 4 and one's digits position is 2. That is, there are 4 tens and 2 ones.

3. Answer: The place value of 43̲ is ones.
Explanation: In a two-digit number the first position is tens digit where the tens number is written and the second position is ones digit where ones number is written.
The given number is 43̲ in which the number 3̲ is underlined. Therefore, the place value of the given number 3̲ is ones.

4. Answer: A
Explanation: Ben has 2 tens (10 + 10 = 20)
The teacher gives 8 ones
(1+1+1+1+1+1+1+1 = 8) So, Ben has 28.

5. Answer: C
Explanation: Given number is 56, 5 tens and 6 ones. 2 tens: 20, 56 − 20 = 36
There are 3 tens and 6 ones = 36.

6. Answer: 25
Explanation: There is 2 tens and 5 ones.
2 tens: 10+10=20, 5 ones: 1+1+1+1+1=5
2 tens + 5 ones = 20+5=25. The number is 25.

ANSWERS AND EXPLANATIONS

7. Answer: The place value of 2̲6 is tens.
Explanation: In a two-digit number the first position is tens digit where the tens number is written and the second position is one's digit where one's number is written.
The given number is 2̲6 in which the number 2 is underlined. Therefore, the place value of the given number 2 is tens.

8. Answer: 7 tens and 9 ones
Explanation: The given number is 79. The number in the tens digit is 7 and ones digits position is 9. That is, there are 7 tens and 9 ones.

9. Answer: 5 tens and 6 ones
Explanation: The given number is 56. The number in the tens digit is 5 and ones digits position is 96. That is, there are 5 tens and 6 ones.

10. Answer: 2 bundles
Explanation: She can make 2 ten.
24 is 2 ten and 4 ones so Sarah can make two groups of 10."

11. Answer: 80
Explanation: There is 8 tens and 0 ones.
8 tens: 10+10+10+10+10+10+10+10 = 80,
0 ones: 0
8 tens + 0 ones= 80 + 0. The number is 80.

12. Answer: A
Explanation: Given 80 cubes.
He gives 5 tens to his partner.
5 tens: 10 + 10 + 10 + 10 + 10 = 50.
80 − 50 = 30 cubes; John has 30 cubes.

13. Answer: C
Explanation: There is 6 tens and 7 ones.
6 tens: 60, 6 ones: 1 + 1 + 1 + 1 + 1 + 1 = 7.
6 tens+7 ones = 60 + 7 = 17. The number is 67.

14. Answer: 110
Explanation: When counting by ones, 110 comes after 109.

15. Answer: A
Explanation: There is 4 rows, with 10 rings in each row: 40 rings. 2 rings in the fifth row. Total: 40 + 2 = 42 rings.

16. Answer: C
Explanation: Given 70 cubes. She gives 3 tens to her partner. 3 tens: 10 + 10 + 10 = 30.
70 − 30 = 40 cubes. Jessy has 40 cubes.

17. Answer: A
Explanation: 6 tens and 3 ones = 60 + 3 = 63.
2 tens and 5 ones = 20 + 5 = 25.
Sum: 63 + 25 = 88. Option A is correct.

18. Answer: 49
Explanations: 6 tens and 9 ones is 69. 20 less than 69 is 49. Rita is thinking of 49.

19. Answer: 31
Explanations: There is 3 tens and 1 ones.
3 tens: 10 + 10 + 10 = 30, 1 ones: 1
3 tens + 1 ones= 30 + 1; The number is 31.

20. Answer: D
Explanation:
Allen has 5 tens (10+10+10+10+10 = 50)
Friend gives 9 ones (1+1+1+1+1+1+1+1+1 = 9)
So, Allen has 59.

4.3 FINDING SUMS AND DIFFERENCES

1. Answer: A
Explanation: James had 76 sunflower stickers and brought 8 more.
So 76+8 = 86 sunflower sticks.

2. Answer: 70
Explanation: Balloons with Sam: 63
He bought 7 more balloons. 63 + 7 = 70

3. Answer: 20 carrots.
Explanation: Total carrots: 50;
Rabbits ate 30 carrots in the morning.
50 − 30 = 20.
The rabbits have 20 carrots left to eat.

ANSWERS AND EXPLANATIONS

4. **Answer: A**
Explanation: Total bananas: 60
Monkeys ate 50 bananas in the morning.
60 − 50 = 10. The monkeys have 10 bananas left to eat.

5. **Answer: B**
Explanation: 45 + 5 = 50.
There are 50 marbles with Miller.

6. **Answer: 40 flowers**
Explanation: 7 tens is 70 and 3 tens is 30.
70 − 30 = 40. Lucy has 40 flowers.

7. **Answer: D**
Explanation: 50 cents + 6 cents = 56 cents.
Ben has 56 cents now.

8. **Answer: C**
Explanation: 38 + 9 = 47
There are 47 blocks with Mike.

9. **Answer: A**
Explanation: 4 + 8 = 12
Alex has 12 boxes of oranges.

10. **Answer: 100 marbles**
Explanation: 6 tens is 60 and 4 tens is 40.
60+ 40 = 100. Jerry has 100 marbles.

11. **Answer: C**
Explanation: 75 cents + 4 cents = 79 cents.
Tim has 79 cents now.

12. **Answer: 6 bundles.**
Explanation: 68 is 6 tens and 8 ones.
Allan can make 6 bundles of 10.

13. **Answer: 4 tens and 3 ones.**
Explanation: 43 is 4 tens and 3 ones.

14. **Answer: 9 bundles.**
Explanation: 91 is 9 tens and 1 ones.
Ronald can make 9 bundles of 10.

15. **Answer: 3 tens and 5 ones.**
Explanation: 35 is 3 tens and 5 ones.

16. **Answer: 9 boxes of candies.**
Explanation: 3 + 6 = 9
Peter has 9 boxes of candies.

17. **Answer: 7 bundles.**
Explanation: 74 is 7 tens and 4 ones.
Donald can make 7 bundles of 10.

18. **Answer: B**
Explanation: 70 − 20 = 50
50 students are still on the playground.

19. **Answer: 2 tens and 8 ones.**
Explanation: 28 is 2 tens and 8 ones.

20. **Answer: 5 bundles.**
Explanation: 53 is 5 tens and 3 ones.
Ben can make 5 bundles of 10.

4.4 CHAPTER REVIEW

1. **Answer: 47 marbles.**
Explanation: There are 10 marbles in first, second, third and fourth row: total 40 flowers. 7 marbles in the fifth row. Total: 40 + 7 = 47 marbles. Silvia has 47 marbles.

2. **Answer: A**
Explanation: The number after 102 is 103.
103 is the answer.Thus, option A Is correct.

3. **Answer: D**
Explanation: 7 tens and 6 ones = 70 + 6 = 76
2 tens and 2 ones = 20 + 2 = 22.
Sum = 76 + 22 = 98. Option D is correct

4. **Answer: 2 tens**
Explanation: There are 13 candies in the first row. There are 12 candies in the second row. Total 13 + 12 = 25, 25 means 2 tens and 5 ones. So, Jerry can make 2 tens.

ANSWERS AND EXPLANATIONS

5. Answer: 108, 109 and 111
Explanation: The number next to 107 is 108.
The number next to 108 is 109.
The number next to 110 is 111.
So the answer is 108, 109 and 111.

6. Answer: 67
Explanation: In the first equation 67 < 68
In the second equation 67 > 66
So, Rita is thinking of number 67, this will make both equations true.

7. Answer: B
Explanation:
5 cartons with 10 ice creams = 50.
7 more ice creams on the table 50 +7 = 57.
There are 57 ice creams on the table.

8. Answer: 46
Explanation: 7 tens and 6 ones = 70 + 6 = 76,
30 less than 76; 76 − 30 = 46.
Mike is thinking of 46.

9. Answer: B
Explanation: There are 47 blocks on the left side and 20 blocks on the right side. 47 is greater than 20, so 47 > 20. So option B is correct.

10. Answer: C
Explanation: 9 tens is 90. She 5 tens and 5 noes to her sister. 5 tens is 50 and 5 ones is 5, 50 + 5 = 55. 90 − 55 = 35.
Mercy has 35 cubes now.

11. Answer: A
Explanation: There are 28 diamonds on the left side and 12 diamonds on the right side. 28 means 2 tens and 8 ones, 12 means 1 tens and 2 ones. 28 + 12 = 40, 40 means 4 tens and 0 ones. So, option A is correct.

12. Answer: 9 tens and 8 ones
Explanation: 98 means 9 tens and 8 ones.

13. Answer: 113, 114 and 116
Explanation: The number next to 112 is 113.
The number next to 113 is 114.
The number next to 115 is 116.
So the answer is 113, 114 and 116.

14. Answer: 2 tens and 9 ones
Explanation: 29 means 2 tens and 9 ones.

15. Answer: B
Explanation: 3 piles and 8 loose stickers = 30 + 8 = 38; He gave away 2 piles and 7 loose stickers = 20 + 7 = 27; 38 − 27 = 11
Edwin has 11 birds stickers with him.

16. Answer: 10 > 8
Explanation: There are 10 butterflies on the left and 8 butterflies on the right sides On comparing, 10 is greater than 8.
The answer is 10 > 8.

17. Answer: 88
Explanation: Four tens means 40. 4 tens and 8 ones = 40 + 8 = 48; 40 + 48 = 88.
The number is 88.

18. Answer: C
Explanation: 2 tens and 1 ones = 20 + 1 = 21
4 tens and 3 ones = 40+ 3 = 43
Sum = 21 + 43 = 64

19. Answer: 100, 101 and 103
Explanation:
99, 100 , 101 ,102 , 103

20. Answer: 76
Explanation: Two tens means 20.
5 tens and 6 ones = 50 + 6 = 56
20 + 56 = 76.
The number is 76.

ANSWERS AND EXPLANATIONS

5. MEASUREMENT

5.1 COMPARE AND MEASURE LENGTHS

1. **Answer: A**
Explanation: Butterfly A is the largest one.

2. **Answer: B**
Explanation: Angel's marker is 10 inches long. Blessy's marker is 5 inches shorter than Angel's marker: 10 − 5= 5 inches. Mercy's marker is 3 inches longer than Angel's marker: 10 + 3 = 13 inches. On comparing the numbers 10, 5 and 13, 5 is the smallest number which represents Blessy's marker.

3. **Answer: D**
Explanation: Plant D is the tallest one.

4. **Answer: Red stick is longer**
Explanation: Red stick length: 30 cm
Blue stick: 15 cm; 30 is larger than 15, so the red stick is longer.

5. **Answer: B**
Explanation: The length of the line is equal to the width of 4 dice.

6. **Answer: Banana, Carrot, and Radish**
Explanation: A carrot of 15 cm, banana of 30 cm and radish is 13 cm. As per the given data, banana, carrot, and radish is the correct Order.

7. **Answer: C**
Explanation: The length of the line is equal to the width of the 5 buttons.

8. **Answer: Pen, ruler, and pencil.**
Explanation: A pen of 12 cm, pencil of 16 cm and ruler is 15 cm. As per the given data, pen, ruler, and pencil is the correct Order.

9. **Answer: 5 inches**
Explanation: First fish: 8 inches, Second fish: 6 inches, Third fish: 3 inches smaller than first = 8 − 3 = 5 inches. The third fish was 5 inches long.

10. **Answer: C**
Explanation: The ball C is the smaller one.

11. **Answer: Pencil is longer**
Explanation: Pencil length: 20 cm, Glue stick: 10 cm, 20 is larger than 10, so pencil's is longer.

12. **Answer: B**
Explanation: Carolina's ribbon: 16 feet. Andrea's ribbon: 4 feet. So Andrea's ribbon is shorter. So, option B is correct.

13. **Answer: Edward, Peter, Jack, and Julie.**
Explanation: Jack is taller than Julie but not as tall as Peter. Edward is taller than Peter. Edward, Peter, Jack, and Julie are in the correct order.

14. **Answer: A**
Explanation: The banana is 4 buttons long

15. **Answer: 3 meter is the smaller one**
Explanation: Given that, 2 cupboards are 3 meters and 6 meters long, respectively. On comparing the numbers 3 and 6, 3 meters is the smaller one.

16. **Answer: Ruler, Pencil, and glue stick**
Explanation: A pencil of 20 cm, ruler of 30 cm and Glue stick is 10 cm. As per the given data, Ruler, Pencil, and glue stick is the correct Order.

17. **Answer: 10 inches**
Explanation: From the figure, the ruler measures the length of the pencil as 10 Inches.

ANSWERS AND EXPLANATIONS

18. Answer: D
Explanation: The butterfly D is the smaller one.

19. Answer: B
Explanation: As per figure, twig B is the shortest one.

20. Answer: B
Explanation: The line is 2 dice long, so the answer is false.

5.2 CLOCK AND TIME

1. Answer: A
Explanation: The digital clock shows Five thirty.

2. Answer: B
Explanation: The hour hand is 2. The minute hand indicates 60 minutes. The time is two o'clock.

3. Answer: A
Explanation: Using numbers, half past three is written as 3:30.

4. Answer: D
Explanation: The hour hand points at 3. Next, read the minutes. The minute hand is pointing to the 6 which means 30 minutes. This denotes 3:30.

5. Answer: B
Explanation: Using numbers, half past four is written as 4:30.

6. Answer: A
Explanation: From the given figure note that, the time in the analog clock is 10:30 and the times in the digital clocks are 10:30, 08:30 and 09:30. So, the matching clock is the first clock.

7. Answer: A
Explanation: A.M. means before 12 noon. P.M. means after 12 noon. The clock is displaying the time 6:38 A.M.

8. Answer: B
Explanation: Using numbers, half past nine is written as 9:30.

9. Answer: B
Explanation: It is after noon so it is P.M., So the time is 01:00 P.M.

10. Answer: Blessy
Explanation: 11:00 is the latest time on the digital clocks.

11. Answer: B
Explanation: Using numbers, half past eleven is written as 11:30. So, it is false.

12. Answer: 08:30
Explanation: The analog clock shows Eight thirty.

13. Answer: A
Explanation: It is morning so it is A.M., So the time is 08:00 A.M

14. Answer: C
Explanation: The hour hand points at 4. Next read the minutes. The minute hand is pointing to the 12 which means 60 minutes. This denotes 4:00.

15. Answer: A
Explanation: Using numbers, half past five is written as 5:30. So, it is true.

16. Answer: A
Explanation: Using numbers, half past ten is written as 10:30. So, it is true.

ANSWERS AND EXPLANATIONS

17. Answer: Workout
Explanation: 7:00 is earlier than 08:00 and 10:00.

18. Answer: D
Explanation: The given time is 12:30 which is half past 12. The time shown in the clock is half past 12 – Option D.

19. Answer: A
Explanation: The hour hand is past the 1 so the hour is 1. The minute hand at the 6 which indicates 30 minutes. The time is 1:30 minutes.

20. Answer: Jack
Explanation: 08:00 is before 09:00. So, Jack finished the homework late.

5.3 CHAPTER REVIEW

1. Answer: A
Explanation: Pencil A is the longest one.

2. Answer: B
Explanation: The hour hand crossed 11, so the hour is 11. Next read the minutes. The minute hand is pointing to the 6 which means 30 minutes. This denotes 11:30.

3. Answer: C
Explanation: Ria's rope is 12 inches long. Maria's rope is 8 inches shorter than Ria's rope: 12 – 8 = 4 inches.
Silvia's rope is 5 inches longer than Ria's rope: 12 + 5 = 17 inches.
On comparing the numbers 12, 4 and 17, 17 is the largest number which represents Silvia's rope.

4. Answer: B
Explanation: The digital clock shows Ten thirty.

5. Answer: D
Explanation: The length of the twig is equal to the width of 3 dice.

6. Answer: A
Explanation: Option A is the smaller one.

7. Answer: AC = 12 km
Explanation: AB = 35 km, BC = 20 km, AC = 12 km Comparing the distances 35, 20, and 12 it can be observed that, the number 12 is the smallest which is the distance between A and C.

8. Answer: A
Explanation: The digital clock shows Two thirty.

9. Answer: A
Explanation: Option A is the larger one.

10. Answer: A
Explanation: From the given figure note that, the time in the analog clock is 10:30 and the time in the digital clocks are 10:30, 09:30 and 10:00. So, the matching clock is the third clock.

11. Answer: B
Explanation: Carolina jog for 7 km. Andrea jog: Distance jogged by Carolina – 2 km = 7 – 2 = 5 km. Andrea jogged less km than Carolina. So, the statement is false.

12. Answer: B
Explanation: Using numbers, half past eleven is 11:30, but it is written as 11:00, so the statement is false.

13. Answer: Angel, Julie, Steffi, Mercy.
Explanation: Steffi is shorter than Julie, but not as short as Mercy. Angel is taller than Julie. Angel, Julie, Steffi, and Mercy is the correct order.

14. Answer: B
Explanation: The time in the given clock is twelve-thirty.

15. Answer: C
Explanation: Team A: 15 points
Team B: Team A − 5 points = 15 − 5 = 10
Team C: Team B + 7 points = 10 + 7 = 17
Team C scored the maximum of 17 points.

16. Answer: D
Explanation: The given time is 1:30 which is half past 1. The time shown in the clock is half past 1 − Option D.

17. Answer: D
Explanation: The length of the line is equal to the width of 3 clips.

18. Answer: B
Explanation: The hour hand shows the 3 and minute hand at the 12 which indicates 3'o clock. The time is 3'o clock.

19. Answer: C
Explanation: Rita sleeps 7 hours a day. Mark sleeps 5 hours a day. Edwin: 3 hours more than mark = 3+5 = 8 hours a day.
So, option C is correct.

20. Answer: B
Explanation: A.M. means before 12 noon. P.M. means after 12 noon.
So, the time on the clock is 04:40 P.M.

6. REPRESENT AND INTERPRET DATA

6.1 ORGANIZE AND REPRESENT DATA IN A CHART

1. Answers: C
Explanation: In the given figure, there are four different groups of items which are zebra, cats, owls and Fishes. Out of the four groups note that, the number of zebras is 7, the number of cats is 5, the number of owls is 8 and the number of fishes is 6. Therefore, the group of owls are largest in number.

2. Answers: B
Explanation: In the given figure, there are four different groups of items which are rings, caps, watches and shoes. Out of the four groups the number of rings is 6, the number of caps is 4, the number of watches is 5 and the number of shoes is 5. Therefore, the group of caps is least in number.

3. Answers: apples-6, grapes-5, and ice-creams-4
Explanation: In the given figure, there are three different eatable things, which are apples, grapes, and ice cream. The tally is the number of items in each group. The number of apples is 6, number of grapes is 5 and the number of ice creams is 4.

4. Answers: B
Explanation: From the figure, note that the number of butterflies is 6.

ANSWERS AND EXPLANATIONS

5. Answers: B
Explanation: Helen bought three different items, which are avocados, apples, and pizza. The number of avocados he bought is 5, the number of apples he bought is 7 and the number of pizzas he bought is 3. Out of the three items, the one he bought the most of the apples.

6. Answers: C
Explanation: The number of spoons he bought is 4, the number of forks he bought is 8 and the number of plates he bought is 2. Out of the three the item he bought the least is the Plates.

7. Answers: B
Explanation: From the given chart, note that there are 10 dogs, 8 cats, and 12 butterflies. Therefore, it is clear that the number of dogs is greater than the number of cats.
The given statement is false.

8. Answers: B
Explanation: From the given chart, note that the vehicles present are 8 cars, 10 trains, and 6 airplanes. Therefore, it is clear that the number of airplanes is the fewest.

9. Answers: B
Explanation: From the given chart, note that the fruits present are 12 apples, 14 grapes and 8 strawberries. Therefore, it is clear that the number of grapes is the largest.

10. Answers: A
Explanation: From the given chart, note that the animals present are 10 bees, 8 grasshoppers, and 6 butterflies. Therefore, it is clear that the number of bees is greater than the number of butterflies.

11. Answers: C
Explanation: There are 3 beetroots; the other groups have more than 3 items.

12. Answers: C
Explanation: There are 7 balls; the other groups have fewer items.

13. Answers: D
Explanation: There are 3 crabs, but the other groups have more items.

14. Answers: A
Explanation: There are 6 flower pots; the other groups have fewer items.

15. Answers: B
Explanation: There are 7 pencils and 6 eraser.

16. Answers: C
Explanation: Max saw 6 gold fishes.

17. Answers: B
Explanation: She saw 7 ladders and 4 rolling chairs. There are fewer rolling chairs.

18. Answers: 𝖨𝖧𝖳 ||
Explanation: There are 7 ice cream popsicles.

19. Answers: 4
Explanation: There are 4 guitars. ||||

20. Answers: The tally and the objects are matched as shown below.
Explanation:

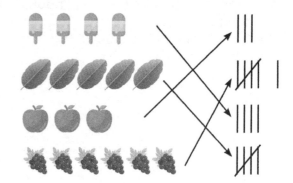

ANSWERS AND EXPLANATIONS

6.2 ORGANIZE AND REPRESENT DATA IN A TABLE

1. Answers: A
Explanation: From the given table, note that the least liked planet is Venus.

2. Answers: A
Explanation: Earth planet is the most popular.

3. Answers: D
Explanation: Total number of books Emily read = 7 + 8 = 15.

4. Answers: B
Explanation: The table clearly shows that Ava purchased a large number of books.

5. Answers: D
Explanation: The table clearly shows the high rainfall in April.

6. Answers: B
Explanation: The sum of the number of candies collected by Polo and Emma is 5 + 7 = 12.

7. Answers: A
Explanation: Amar has only 2 candies. So, Amar has the least compared to others.

8. Answers: C
Explanation: Nochu ate the most donuts when compared to the others.

9. Answers: B
Explanation: According to the table, May had more holidays compared to other months.

10. Answers: B
Explanation: From the table, Jack has the least amount of candy. As a result, the statement is false.

11. Answers: A
Explanation: The number of students who preferred bicycles over buses is 5 − 4 = 1.

12. Answers: C

| Bikes | 3 |
| Cars | 6 |

Explanation: Number of bikes 3 and cars 6. 3 + 6 = 9.

13. Answer: Apples-5; Ice-creams- 6; Popcorns- 4
Explanation:

Apples	6
Icecreams	5
Popcorns	4

14. Answers: Stars-6 , Hearts-5, and triangles-4.
Explanation:

Stars	卌 I
Hearts	卌
Triangles	IIII

15. Answers: B
Explanation: From the table, note that, the number of Audi cars sold on Tuesday is 12 and the number of BMW cars is 28. Also note that on Wednesday, the number of cars sold for the Audi was 73. Therefore, it is clear that Tuesday is not the day of the highest sales for all the brands.

16. Answers: D
Explanation: From the given table, note that, the number of nails left on Friday is 4 and the number of nuts left is 14. Therefore, the number of nuts more than nails is, 14 − 4 = 10.

17. Answers: A
Explanation: From the given chart, note that there are 2 jackets, 4 pants, and 8 shirts. Therefore, it is clear that the number of shirts is more than the other two.

ANSWERS AND EXPLANATIONS

18. Answers: B
Explanation: From the given table, note that the number of cones sold on Friday is 4, Saturday is 5, and Sunday is 3. Thus, the total number of Sundae cones sold on these three days is 4+5+3 = 12.

19. Answers: A
Explanation: From the given table note that Emma had 9 pink lilies, 6 Red lilies, and 8 yellow lilies. Therefore, it is clear that pink Lilly is Emma's favorite.

20. Answers: B
Explanation: From the above table, it can be noted that there are 4 avocados, 5 apples, 6 grapes, and 5 pears. Therefore, it is clear that avocado is the least liked fruit.

6.3 INTERPRET DATA WITH UP TO FOUR CATEGORIES

1. Answers: A
Explanation: The bar for Wednesday shows 5 pages.

2. Answers: B
Explanation: Utah had the best gas prices - $2 a gallon.

3. Answers: A
Explanation: The height of the bar graph is highest on Wednesday.

4. Answers: Graph 2
Explanation: The heights of the bars in graph 2 match the numbers in the problem.

5. Answers: B
Explanation: She brought 6 apples.

6. Answers: A
Explanation: From the bar graph given, note that there are 3 different categories of fruits which are Oranges, Apples, and Strawberries. It can be noted that the number of students who like strawberry is 6.

7. Answers: 10
Explanation: From the given chart it can be noted that three different colors of leaves are collected which are green, red, and yellow. From the chart it can be noted that there are 4 red leaves and 6 yellow leaves. Therefore, the total number of yellow and red leaves is 6 + 4 = 10.

8. Answers: C
Explanation: The given chart, note that there are three different fruits, which are strawberries, grapes, and apples. Also, note that 4 people like grapes, and 7 people like apples. Therefore, the number of apples more than grapes is, 7 - 4 = 3.

9. Answers: B
Explanation: From the given pie chart, note the number of people who like toast is 30, number of people who like pancakes is 50 and the number of people who like to eat eggs is 20. Therefore, in total the number of people who preferred eggs than pancake is 50 - 20 = 30.

10. Answers: A
Explanation: According to the tally numbers, note that, the number of strawberries is 7, number of cookies and creams is 4 and the number of fudges is 3. Out of the three, it can be noted that strawberries are the most preferred.

11. Answers: C
Explanation: From the given chart, note that, there are three different fruits, which are strawberries, cookies and creams, and fudge. Also, note that 7 people like strawberries, 4 people like cookies and creams, and 3 people like fudge. Therefore, it is clear that fudge are the least preferred.

ANSWERS AND EXPLANATIONS

12. Answers: A
Explanation: From the given chart, note that there are three different subjects, which are English, math, and science. In addition, 5 students preferred English, 7 preferred math, and 4 preferred science.

13. Answer: B
Explanation: From the given figure, note that, there are three different flavors of candies, which include apple, orange, and kiwi. Out of the three flavors, note that 9 students like the apple flavor, 6 students like the orange flavor, and 4 students like the kiwi flavor.

14. Answers:

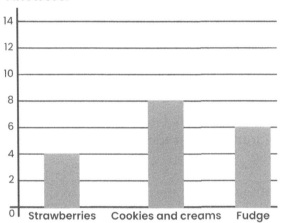

Explanation: According to the tally numbers, the number of strawberries is 4, number of cookies and creams is 8, and the number of fudges is 6. The tally numbers are colored as shown in the blocks.

15. Answers: B
Explanation: From the bar graph given, note that there are 3 different categories of fruit, which are oranges, apples, and strawberries. It can be noted that the number of students who like strawberries is 5, the number of students who like oranges is 4 and the number of students who like apples is 7. Out of the three, it is clear that the number of apples is the largest.

16. Answers: A
Explanation: From the bar graph given, note that there are 3 different categories of fruits, which are oranges, apples, and strawberries. It can be noted that the number of students who like strawberry is 5

17. Answers: B
Explanation: From the given figure, note that there are three different foods, which include burgers, pizza, and sandwiches. Out of the three foods, note that 8 students like the burger, 7 students like the pizza, and 9 students like the sandwich.

18. Answers: C
Explanation: From the given chart, note that, there are three different vegetables, which are cucumber, beet, and broccoli. Also, note that 6 people like cucumbers, 7 people like beets, and 8 people like broccoli. Therefore, it is clear that broccoli is the most preferred.

19. Answers: B
Explanation: From the given figure note that, there are three different flavors of candies which includes banana, grapes, and orange. Out of the three flavors note that 4 students like banana flavor, 5 students like orange flavor, and 6 students like grapes.

20. Answers:

	Strawberries	Cookies	Cakes
8			
7	■		
6	■		
5	■	■	
4	■	■	■
3	■	■	■
2	■	■	■
1	■	■	■

Explanation: According to the tally numbers, note that, the number of strawberries is 7, number of cookies is 5 and the number of fudges is 4. The tally numbers are shaded as shown in the blocks.

ANSWERS AND EXPLANATIONS

6.4 CHAPTER REVIEW

1. **Answers: A**
Explanation: There are only 2 pineapples. The other groups are more than 2 items.

2. **Answers: A**
Explanation: There are 5 apples, 6 avocados, 4 grapes, and 5 pears.

3. **Answers: B**
Explanation: Mary had 5 trophies.

4. **Answers: A**
Explanation: The bar for Monday is the highest.

5. **Answers: A**
Explanation: The bars in graph A match the numbers in the problem.

6. **Answers: B**
Explanation: He found 7 watches and 5 bottles. 5 is less than 7.

7. **Answers: A**
Explanation: The bar graph shows she drove 10 miles on Sunday

8. **Answers: 8**
Explanation: From the given chart, it can be noted that three different colors of leaves are collected, which are green, red, and yellow. From the chart, it can be noted that there are 2 green leaves and 6 yellow leaves. Therefore, the total number of yellow and green leaves is 6+2=8.

9. **Answers: B**
Explanation: Kisha ate the most donuts when compared to the others.

10. **Answers: D**
Explanation: From the table, August had the fewest holidays compared to other months.

11. **Answers: C**
Explanation: In comparison to other modes of transportation, bikes are used by the majority of students.

12. **Answers: A**
Explanation: From the given chart, Ice creams - 4 and rabbits - 4 both in the same group.

13. **Answer: D**
Explanation: There are 8 hearts.

14. **Answers: B**
Explanation: Total number of students 12 + 14 + 20 = 46.

15. **Answers: A**
Explanation: From the given chart, the number of tomatoes more than cabbages is, 8 – 4 = 4.

16. **Answers: C**
Explanation: 12 students liked the orange flavor.

17. **Answers: B**
Explanation: The number of apples in a week 1 is 4 and a week 3 is 3. Total number of apples in week 1 and 3 is 4 + 3 = 7.

18. **Answers: B**
Explanation: There are five alligators.

19. **Answers: B**
Explanation: Math is a favorite subject among students.

20. **Answers: C**
Explanation: From the bar graph, there are 9 drum toys were sold.

ANSWERS AND EXPLANATIONS

7. GEOMETRY

7.1 GEOMETRIC PROPERTIES – 2D AND 3D SHAPES

1. **Answer:** Answer will vary
Explanation: The shape below is a diamond.

2. **Answer:** Circle
Explanation: The shape below has a curved surface.

The above figure shows a closed shape with three sides.

3. **Answer:** Triangle

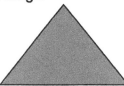

Explanation: The above figure shows a closed shape with three sides.

4. **Answer:** The shape is a circle.

Explanation: A shape with 0 vertices and 0 sides is a circle.

5. **Answer:** The given shape is closed.
Explanation: Note that the given figure is a closed square.

6. **Answer:** There are 5 sides.
Explanation: The given figure, note that, there are 5 sides.

7. **Answer:** The shape is a cube.
Explanation: A shape with squares on all 6 faces is a cube.

8. **Answer:** A
Explanation: The given shape is an open.

9. **Answer:** Zero vertices.
Explanation: The given ball shape is a circle, and it has zero vertices.

10. **Answer:** The given shape is closed.
Explanation: From the given figure, note that the shape is closed.

11. **Answer:** 8 vertices
Explanation: In a cube, there are 8 vertices in total.

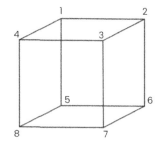

12. **Answer:** The shape is a cylinder.
Explanation: A shape with curved surfaces and circular faces at the top and bottom is a cylinder.

13. **Answer:** Oval
Explanation: The given shape is Oval.

14. **Answer:** 8 sides.
Explanation: The given shape has 8 sides.

15. **Answer:** C
Explanation: The second shape is open and all other 3 shapes are closed.

ANSWERS AND EXPLANATIONS

16. Answer: The given shape is a cylinder.
Explanation: The shape has 2 faces, 0 vertices, and 0 edges. Therefore, the shape is a cylinder.

17. Answer: Answer will vary

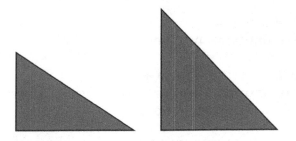

Explanation: The above figure shows a triangle shape in two different sizes.

18. Answer: Answer will vary
Explanation:

19. Answer: There are 6 edges.
Explanation: The given shape has 6 edges.

20. Answer: The statement is False.
Explanation: The shape has 5 vertices and is a Pentagon.

7.2 IDENTIFY AND DRAW 2D AND 3D SHAPES

1. Answer:

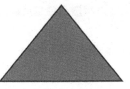

Explanation: It is known that a triangle is 2D shape with vertices and three sides.

2. Answer:

Explanation: It is known that a diamond is a 2D shape with 4 vertices and 4 sides.

3. Answer:

Explanation: The required figure with the square inside the circle is obtained as shown above.

4. Answer:

Explanation: From the figure, note that Figure A is a circle and Figure B is a triangle.

5. Answer: A triangle has 3 sides.
Explanation: A triangle is a 2D shape with three sides and three vertices.

ANSWERS AND EXPLANATIONS

6. Answer:

The name of the figure is a rectangle.
Explanation: A rectangle is a 2D shape with four sides and four vertices.

7. Answer: A heptagon has 7 sides.
Explanation: A heptagon is a 2D shape with seven sides and seven vertices.

8. Answer:

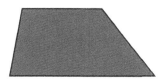

Explanation: A trapezoid is a 2D shape with 4 sides and 4 vertices.

9. Answer: Figure B
Explanation: Figure A is a rectangle which is a 2D shape with 4 sides and 4 vertices and Figure B is a triangle with 3 vertices and three sides. Therefore, the shape with fewer vertices is a triangle.

10. Answer: B
Explanation: The shape in the figure is a complete triangle which is 2D shape with 3 sides and 3 vertices. Therefore, the given shape is closed.

11. Answer:

Explanation: A kite is a 2D shape with 4 vertices and 4 sides.

12. Answer:
Explanation:
The required figure is.

13. Answer:

Explanation: The shape of the square is 3D which is not true. A square is a 2D shape with 4 sides and 4 vertices.

14. Answer: B
Explanation: The shape of the square is 3D which is not true. A square is a 2D shape with 4 sides and 4 vertices.

15. Answer: Figure B
Explanation: The first shape is a circle, which is a 2D shape has 0 sides and the second shape is a square, which is also a 2D shape with 4 sides and 4 vertices. Therefore, the shape with the most sides is the square.

16. Answer: Figure A
Explanation: From the figure, note that the first shape is a triangle, which is a 2D shape with three sides and three vertices; the second shape is a square, which is a 2D shape with four sides and four vertices; and the third shape is a pentagon with 5 vertices and 5 sides.

17. Answer: A triangular prism has 9 edges.
Explanation: A rectangular prism is a 3D image with 6 vertices, 5 faces, and 9 edges. Therefore, a rectangular prism will have 9 edges.

18. Answer: B
Explanation: A sphere is a 3D shape with no corners and edges, whereas a rectangular prism is a 3D shape with 12 edges, 6 faces, and 8 corners

ANSWERS AND EXPLANATIONS

19. Answer: The given shape is 3D.
Explanation: There is a 1 face, but no edges or vertices. Therefore, the given shape is 3D.

20. Answer: The shape is a cube.
Explanation: There are 8 vertices, 6 square faces, and 12 edges. Therefore, the given shape is a cube.

> **7.3 CREATE COMPOSITE FIGURES IN TWO OR THREE DIMENSIONS**

1. Answer: The shapes A and C will make a square.
Explanation: Note that if the shapes A and C are combined, a rectangle will be formed.

2. Answer: The four small squares are combined to form a bigger square.
Explanation:

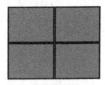

3. Answer:

Explanation: The shape is made up of two triangles.

4. Answer:

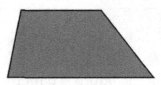

Explanation:
A shape with 4 sides, four corners with the length of one side short and the other side long is a trapezoid.

5. Answer:

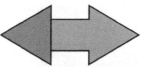

Explanation: The required shape is as shown above.

6. Answer:

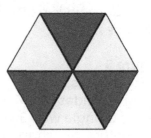

Explanation: From the above figure, note that the hexagon is formed using the shapes of 6 triangles.

7. Answer:

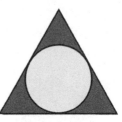

Explanation: Given that, the shape of the gift is formed by using a shape with three sides and three corners and then inscribing a shape inside it with no edges or vertices. Thus, the required shape is as shown above.

8. Answer: B
Explanation: The two figures are 2 Dimensional.

ANSWERS AND EXPLANATIONS

9. Answer: A kite shape will be formed.
Explanation: If a triangle and an inverted triangle are combined, a kite shape will be formed.

10. Answer:
Linda made a shape with 5 edges and 5 corners. The required shape will be a pentagon.

Explanation:

11. Answer: Cylinder and a cone
Explanation: From the figure, note that a cylinder and a cone are used to make the required 3D shape.

12. Answer: B
Explanation: If two squares are joined, they form a rectangle.

13. Answer: Cylinder
Explanation: The 3D shapes used in the given figure are a Cylinder.

14. Answer: Rectangular prism.
Explanation: The 3D shapes used in the given figure are a rectangular prism.

15. Answer: The shape similar to a carrot is the cone.
Explanation: The carrots' similar shape is a cone.

16. Answer: The 3D shapes used are sphere and cone.
Explanation: The 3D shapes used in the given figure are a sphere and a cone.

17. Answer: The 3D shapes are cone and cylinder.
Explanation: The 3D shapes in the given figure are a cone and a cylinder.

18. Answer: Rectangle and triangle.
Explanation: You can make a rectangle by splitting the square either vertically or horizontally. You can make a triangle when you split the square diagonally.

19. Answer: Given that, Ava made a shape with 3 edges and 3 corners. Therefore, the required shape will be a triangle.
Explanation:

20. Answer: No, the given figure doesn't have a curved surface.
Explanation: From the given figure, it can be observed that there is no curved surface.

7.4 CHAPTER REVIEW

1. **Answer: The shapes used are triangle and a square.**
Explanation: A triangle and a square are used to make the required shape.

2. **Answer: D**
Explanation: You make triangle when you fold a square piece of paper diagonally.

ANSWERS AND EXPLANATIONS

3. Answer: A
Explanation: A square has 4 equal sides.

4. Answer: Answer will vary
Explanation: The shape is made up of two triangles.

5. Answer: C
Explanation: A circle is a shape with 0 corners and 0 sides.

6. Answer: D
Explanation: A square that is split diagonally makes a triangle.

7. Answer: D
Explanation: A basketball is a sphere. It is spherical in shape.

8. Answer: C
Explanation: A rhombus is diamond-shaped. It has four equal sides as well as four corners.

9. Answer: 6
Explanation: A hexagon has 6 vertices.

10. Answer:

Explanation: A 2D pentagon has five sides.

11. Answer: The given shape is 2D.
Explanation: From the given figure, note that there are no faces, no edges, and no vertices. Therefore, the given shape is 2D.

12. Answer: 5
Explanation: A given shape has 5 sides

13.

Explanation: The first shape is a circle, which is a 2D shape and the second shape is a triangle, which is also a 2D shape with 3 sides and 3 vertices. Therefore, the shape with the most sides is the triangle.

14. Answer: B
Explanation: The shape of the pentagon is 3D which is not true. It is known that a Pentagon is a 2D shape with 5 sides and 5 vertices.

15. Answer: B
Explanation: The shape in the figure is a complete circle, which is a 2D shape with no sides and no vertices. Therefore, the given shape is closed.

16. The given shape is a triangle.
Explanation: The shape has with 3 faces, 3 edges, and 3 vertices.
Therefore, the shape is a triangle.

17. Answer:

Explanation: The above figure shows a closed shape with sides.

18. Answer: 3
Explanation: A triangle has 3 vertices.

19. Answer: A
Explanation: The give figure is a sphere, so it has a curved surface.

20. There are no 2D shapes used in the figure.
Explanation: In the given figure, no 2D shapes are used, only the cone and cylinder are used, which are 3D shapes.

ANSWERS AND EXPLANATIONS

8. PARTITION SHAPES

8.1 DIVIDE CIRCLES AND RECTANGLES INTO TWO OR FOUR EQUAL PARTS

1. **Answer: A**
Explanation: Shape A shows equal parts.

2. **Answer: C**
Explanation: Shape C shows equal parts.

3. **Answer: A**
Explanation: Halves means 2 equal parts.

4. **Answer: B**
Explanation: Fourths means 2 equal parts.

5. **Answer: A**
Explanation: 1 out of 4 equal parts is shaded.

6. **Answer: A**
Explanation: Yes, a given watermelon is cut in half evenly.

7. **Answer:**

Explanation: The circle is partitioned into two equal halves as shown above.

8. **Answer: 4**
Explanation: A rectangle shape picture is partitioned into four equal parts.

9. **Answer: Answer will vary**

Explanation: The lemon is partitioned into two equal parts.

10. **Answer:**

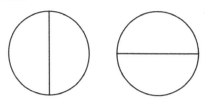

Explanation: The given figure is divided into two equal halves in two different ways.

11. **Answer:**

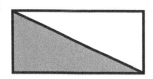

Explanation: There are two partitions. Therefore half of the figure is shaded as shown above.

12. **Answer: The given statement is true.**
Explanation: From the given figure, note that the rectangle is divided into four equal parts.

13. **Answer: The given statement is true.**
Explanation: The circle is divided into four equal parts.

14. **Answer: True**
Explanation: The picture is divided into 2 equal parts.

15. **Answer: False**
Explanation: The picture is divided into 4 equal parts.

16. **Answer:**

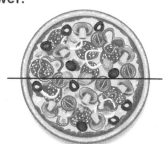

Explanation: The given picture is divided into two equal parts.

305

ANSWERS AND EXPLANATIONS

17. Answer: True
Explanation: The picture is divided into 2 equal parts.

18. Answer:

Explanation: The given shape is divided into two equal parts.

19. Answer: B
Explanation: The circle is divided into 4 equal parts.

20. Answer: Four
Explanation: The dartboard is divided into 4 equal parts.

8.2 DESCRIBE SHARES AS HALVES, FOURTHS, AND QUARTERS

1. Answer: No
Explanation: The given shape is not divided into halves.

2. Answer: Yes
Explanation: The given shape is divided into fourths.

3. Answer:

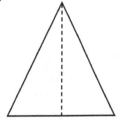

Explanation: The given shape is divided into halves.

4. Answer:

Explanation: The second shape, which is the star, is divided into fourths.

5. Answer:

Explanation: The given figure is divided into 2 parts. Therefore, coloring half means one part is colored.

6. Answer: True
Explanation: The oval is divided into halves.

7. Answer: B
Explanation: In the figure, the whole shape is shaded.

8. Answer: 4
Explanation: The given shape is divided into 4 parts.

9. Answer:

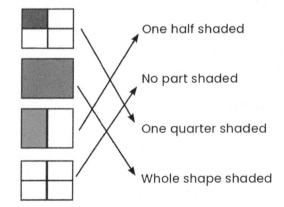

306

ANSWERS AND EXPLANATIONS

10. Answer: A
Explanation: One half of the shape is colored

11. Answer: Half part is unshaded.
Explanation: It can be observed that half of the figure is unshaded.

12. Answer: 2
Explanation: Half of the parts are shaded.

13. Answer:

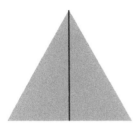

Explanation: The triangle shape is divided into less parts.

14. Answer: One part
Explanation: The one part is colored.

15. Answer: The given shape is divided into quarters.
Explanation: The given shape is divided into four.

16. Answer: Fourths
Explanation:

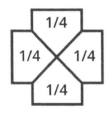

17. Answer: Each part is labeled 1/2
Explanation:

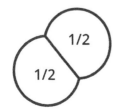

18. Answer: The given shape is divided into quarters.
Explanation: The given shape is divided into four, which means quarters.

19. Answer: 4
Explanation: The given shape is divided into 4 parts.

20. Answer: No
Explanation: The shape is not divided into halves.

8.3. IDENTIFY EQUAL AND UNEQUAL SHARES

1. Answer: The first heart is equally divided.
Explanation: There are two hearts in which the first heart is equally divided, and the second heart is not. Thus, the first shape shows equal shares.

2. Answer: The second oval is unequally divided.
Explanation: There are two ovals, the first of which is equally divided, and the second of which is unequally divided. Thus, the second shape shows unequal shares.

3. Answer: The first shape and the third shape show the equal shares.
Explanation: The first and third are equally divided and the second and fourth of which are unequally divided. Thus, the first shape and the third shape show equal shares.

4. Answer: A
Explanation: The given shape is a parallelogram, which is equally shared.

5. Answer: B
Explanation: From the given figure, note that the rectangle is unequally divided.

ANSWERS AND EXPLANATIONS

6. Answer: B
Explanation: From the given figure, note that the rectangle is unequally divided.

7. Answer: The first shape shows equal shares.
Explanation: From the given figure note that, there are two triangles in which the first is equally divided and the second is unequally divided. Thus, the first shape shows equal shares.

8. Answer:

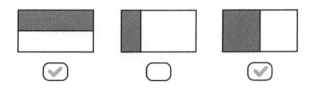

Explanation: The shape with equal shares is marked.

9. Answer: A

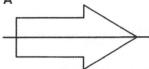

Explanation: Yes, it is possible to divide the set into equal shares.

10. Answer:

Explanation: The given shape is divided into equal parts.

11. Answer:

Explanation: The given shape is divided into two equal shares

12. Answer: B
Explanation: No, it is not possible to divide the set into equal shares.

13. Answer:

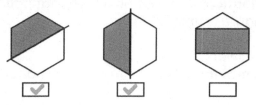

Explanation: The shape with equal shares is marked.

14. Answer:

Explanation: The shape with unequal shares is marked.

15. Answer: B
Explanation: The given is divided into unequal shares.

16. Answer: Yes
Explanation: Yes, it is possible to divide into equal shares.

17. Answer: Yes
Explanation: Yes, it is possible to divide into equal shares.

18. Answer:

Explanation: The shape with unequal shares is marked.

ANSWERS AND EXPLANATIONS

19. Answer:
Explanation: The shape with equal shares is marked.

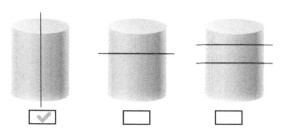

20. Answer: A
Explanation: The given shape is an oval, which is equally shared.

8.4 CHAPTER REVIEW

1. Answer: A
Explanation: Halves means 2 equal parts.

2. Answer: B
Explanation: Fourths means 4 equal parts.

3. Answer: B
Explanation: The rectangle is divided unequally.

4. Answer: B
Explanation: It can be observed that the given picture is partitioned into two equal parts.

5. Answer: C
Explanation: It can be observed that the circle is partitioned into two equal parts.

6. Answer: A
Explanation: The first picture is divided into equal shares.

7. Answer: D
Explanation: The whole shape is shaded.

8. Answer: B
Explanation: Fourths means 4 equal parts.

9. Answer: A
Explanation: Yes, a given candy is cut in half evenly.

10. Answer: 2
Explanation: From the given figure it can be observed that the given picture is partitioned into two equal parts.

11. Answer:

Explanation: The tomato is partitioned into two equal parts.

12. Answer:

Explanation: In the given figure, there are two partitions. Therefore half of the figure is shaded as shown above.

13. Answer:

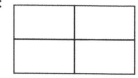

Explanation: The given rectangle is divided into 4 equal parts.

14. Answer: A
Explanation: Yes, the given shape is divided into halves.

15. Answer: No
Explanation: The given shape is not divided into halves.

16. Answer: A
Explanation: One half of the shape is colored

ANSWERS AND EXPLANATIONS

17. Answer: The shape is divided into quarters.
Explanation: The shape is divided into four which means quarters.

18. Answer: No
Explanation: The given shape is not divided into fourths.

19. Answer:

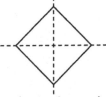

Explanation: The given shape is divided into quarters.

20. Answer: B
Explanation: In the figure, the whole shape is shaded.

COMPREHENSIVE ASSESSMENT

ASSESSMENT – 1

1. Answer: A
Explanation: Heart stickers with Rita: 35
Her friend gave her 20 more.
35 + 20 = 55.
Rita has 55 heart stickers altogether.

2. Answer: D
Explanation: Total is the number of 12, 7 flowers on the left side.
12 − 7 = 5, it can also be written as 7 + 5 = 12, 5 is the missing number.

3. Answer: C
Explanation: Tim: 5 bikes; Jessica: 9 bikes
Riya: 6 bikes. 5 + 9 + 6 =20
Totally, there are 20 bikes.

4. Answer: 9
Explanation: 9 flowers on the first row and 6 flowers on the second row. 9 + 6 = 15, So 9 is the missing number.

5. Answer: B
Explanation: Practice in this month: 12 days
Practice in next month: 7 days; 12 + 7 = 19
Total: 19 days.

6. Answer: A
Explanation: You can use the Associative property of addition to combine the 7 and 3 to make 10.

7. Answer: 11 paintings
Explanation: Sarah: 7 flower paintings
Benny: 4 flower paintings; 7 + 4 = 11.
In total, there are 11 flower paintings.

8. Answer: C
Explanation: Red rose : 8; Pink rose: 5
Yellow rose: 4; The equation can be written as 8 + 5 + 4 =?. So, option C is correct and their total is 8 + 5 + 4 = 17.

9. Answer: 5
Explanation: 14 − 9 = 5. It can also be written as 14 − 5 = 9; So the missing number is 5.

10. Answer: C
Explanation: Strawberries: 14; Apples: 7
14 − 7 = 7, So, there are 7 more strawberries than apples.

11. Answer: D
Explanation: Max has drawn 13 cupcakes. He now needs to cross out 7 cupcakes to represent the 7 cupcakes that were eaten. 13 − 7 = 6; 6 cupcakes were left.

12. Answer: 7 burgers
Explanation: Total burgers: 13; Franklin brought 6 burgers. 13 − 6 = 7. His mom gave him 7 burgers.

13. Answer: D
Explanation: Total apples with Tom: 12
He gave 4 apples to his friend. 12 − 4 = 8, the number 8 can be written as 10 − 2 = 8
Tom gives 8 apples to his sister.

ANSWERS AND EXPLANATIONS

14. Answer: D
Explanation: There are 5 blocks.

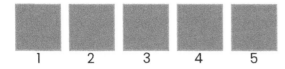

15. Answer: 14 dresses.
Explanation: There are 9 golden dresses + 5 Silver dresses: 14. It can also be expressed as 5 Silver dresses + 9 golden dresses = 14. So, the answer is 9 + 5 = 14, 5 + 9 = 14. Steffi has 14 dresses in total.

16. Answer: A.
Explanation: Yes. A given diagram is cut equally in half.

17. Answer: B
Explanation: First expression: 18 − 3 = 15. Second expression: 19 − 4 = 15. Both are equal, so Tim is right.

18. Answer: 6 + 3 = 4 + 5.
Explanation: In Figure A 6 + 3 Pepper = 9 Pepper. In figure B 4 + 5 Pepper = 9 Pepper in the right side. Both sides are equal. So, the answer is 6 + 3 = 4 + 5.

19. Answer: 10 + 5 = 15.
Explanation: There are 8 burgers, 5 hotdogs, and 2 pizzas. 8 + 5 + 2 = 15, It can also be written as 10+5=15. There are 15 items in the party.

20. Answer: 11 plants
Explanation: Red kites: 8; Blue kites: 6
3 kites flew away. 8 + 6 - 3 = 11.
Joseph has 11 plants

21. Answer: C
Explanation: The number next to 108 is 109. So she will count 109. Thus, option C is correct.

22. Answer: Agree.
Explanation: I agree because 25 questions is 10 more than 15. So, the equation is correct.

23. Answer: A
Explanation: 89 means 8 tens and 9 ones. So option A is correct.

24. Answer: B
Explanation:
3 stick of ten cubes = 10 + 10 + 10 = 30
She gets 9 more cubes. 30 + 9 = 39
Jessica has 39 cubes in all.

25. Answer: D
Explanation:
Green apples: 37 is split into 30 + 7,
Mangoes: 23 is split into 20 + 3. The total would be represented by 30 + 7 + 20 + 3.

26. Answer: D
Explanation:
37 empty pages in the first notebook.
10 empty pages in the second notebook.
12 empty pages in the third notebook.
37 + 10 + 12 = 59. 59 means 5 tens and 9 ones. Option D is correct.

27. Answer: 2 tens.
Explanation: Totally, there are 25 flowers. 25 is 2 tens and 5 ones.
So, Jenny can make 2 tens.

28. Answer: C
Explanation: The hour hand crossed 2 so it's 2 hours. Next, read the minutes. The minute hand is pointing to the 6 which means 30 minutes. This denotes 02:30.

29. Answer: 34
Explanation: 5 tens and 9 ones = 50 + 9 = 59
2 tens and 7 ones = 20 + 7 = 27; 59 − 27 = 32
32 means 3 tens and 2 ones.

ANSWERS AND EXPLANATIONS

30. Answer: B
Explanation: The length of the line is equal to the width of four dice.

31. Answer: B
Explanation: Catherina's classroom: 15 feet. Bethel's classroom: 3 feet long + Catherina's classroom. 3 + 15 = 18 As a result, Bethel's classroom is longer. So, option B is correct.

32. Answer: 08:30
Explanation: The time on the given clock is 8:00. It is given that the clock is 30 minutes slow. Therefore, the correct time is 08:00 + 00.30 = 08:30.

33. Answer: 10
Explanation: 8 + 2 = 10

34. Answer: B
Explanation: 12 + 6 = 18, which can also be written as 18 - 7 = 11; 11 is the missing number.

35. Answer: A
Explanation: The given rectangle is divided into 6 parts. 3 out of 6 parts are not shaded.

36. Answer: 9
Explanation: 5 + 4 = 9.

37. Answer: D
Explanation: Cupcakes with Ava: 9; Cupcakes from her friend: 6; 9 + 6 = 15 Ava has 15 cupcakes.

38. Answer: 4
Explanation: The expression 3 + 3 + 8 = 14. It can be written as 10 + 4 = 14. Therefore, 4 + 3 + 8 = 10 + 4.

39. Answer: B
Explanation: Sketches with John: 10 He gave some to his brother. Remaining sketches with John : 5 10 − 5 = 5; The missing number is 5.

40. Answers: A
Explanation: According to the tally numbers, note that, the number of burgers is 7, number of pizzas is 4 and the number of hotdogs is 3. Out of the three it can be noted that Strawberries is the most preferred one.

41. Answers: Both shapes are 2D and closed.
Explanation: From the given figure, note that both figures are 2D and closed.

42. Answer: B
Explanation: A rectangle has 4 sides with opposite sides equal.

43. Answer: C
Explanation: Add the number of oranges bought and the number of oranges received to find how many there are in all. 24 + 7 = 31.

44. Answer: B
Explanation: The shape with the most vertices is a hexagon.

45. Answer: A
Explanation: Breakfast time is in the morning, before 12:00 P.M. So it is A.M.; the time is 08:30 A.M. So it is true.

COMPREHENSIVE ASSESSMENT

ASSESSMENT – 2

1. Answer: A
Explanations: Alice bought 8 ice creams. Her friend gave 9 more 8 + 9 = 17. Alice has 17 ice creams in total.

2. Answer: C
Explanations: 11 + 9 = 20. 20 is the missing number.

ANSWERS AND EXPLANATIONS

3. Answer: B
Explanations: Monday: 7 pictures; Tuesday: 4 pictures; Wednesday: 3 pictures. 7 + 4 + 3 = 14. Mercy draws 14 pictures altogether.

4. Answer: 8
Explanation: 5 + 9 = 14, 6 is the number on the right side. 14 − 6 = 8. 5 + 9 = 14 can also be written as 8 + 6 = 14; So, 8 is the missing number.

5. Answer: 10 + 5 = 15 fruits in total.
Explanation: 7 Green apples + 3 oranges + 5 strawberries = 7 +3 + 5 = 15. The number 15 can be written as 10 + 5 = 15. So, 7 + 3 + 5 = 15 can be written as 10 + 5 = 15

6. Answer: D
Explanation: Lisa: 7 photo frames; Ellen: 3 more frames than Lisa = 3 + 7 = 10. Total: 7 + 10 = 17. There are 17 photo frames altogether.

7. Answer: A
Explanation: Chocolates with Lisa: 29 She distributed 19 chocolates 29 − 19 = 10, Taking away 19 from 29 gives 10. 10 chocolates are left.

8. Answer: 16 marbles.
Explanation: Red marbles: 7; Blue marbles: 5 Pink marbles: 4; 7 + 5 + 4 = 16. The number 16 can be written as 10 + 6. Total: 10 + 6 = 16, Rita has 16 marbles with her.

9. Answer: |||| ||
Explanation: The number of butterflies is 7.

10. Answer: D
Explanation: There are 14 flowers total with 4 of them separate. You can represent this as 14 - 4; 14 − 4 = 10.

11. Answer: B
Explanation: Total vegetables: 15; Broccolis: 4 Carrots: 5; 15 − 4 − 5 = 6, Take away 9 from 15 to give 6. So, there were 6 Pepper.

12. Answer: 2
Explanation: Explanations will vary but must include "He subtracted 4 from 14 to get 10 and now needs to subtract 2." 10 − 2 = 8, 2 will be in the blank, and the answer is 8.

13. Answer: A
Explanation: Andrew has 15 bike toys. He gave 8 toys to his friend. 15 − 8 = 7, Andrew has 7 bike toys left with him.

14. Answer: 5
Explanation: Total Pepper: 12; Takeaway: 7; 12 − 7 = 5 , takeaway 7 from 12 gives 5; Related sentence is 12 − 5 = 7; 5 is the missing number.

15. Answer: D
Explanation: Given expression: 7 + 5 = 12 It only matches with the expression 12 = 5 + 7 Thus the addition sentence 7 + 5 = 12 matches with 12 = 5 + 7.

16. Answer: 14 flowers
Explanation: Angel has 3 lilies + 7 roses + 4 lotus = 14 flowers 3 + 7 + 4 = 14 It can also be written as 10 + 4 = 14. So, the answer is 3 + 7 + 4 = 10 + 4 = 14. Angel has 14 flowers.

17. Answer: C
Explanation: Black jeans: 5; Blue jeans: 6; 5 + 6 = 11, The number 11 can be written as 10+1; So, 5 + 6 = 11 = 10 + 1. David buys 11 jeans in total.

18. Answer: A
Explanation: Here 9 + 5 = 14, It also matches with 8 + 6 = 14, Both values are the same. So, option A is true.

ANSWERS AND EXPLANATIONS

19. Answer: D
Explanation: Here, 7 + 7 = 14 and 17 − 3 = 14
So, the only expression 7 + 7 is equal to 17 − 3
So, option D is correct.

20. Answer: A
Explanation: There are 6 dots + 3 dots = 9 dots on the left side; There are 5 dots + 4 dots = 9 dots on the right side. 6 + 3 = 9 is equal to 5 + 4 = 9, So it is True.

21. Answer: 16 items.
Explanation: Rita bought 8 sandwiches, 5 hotdogs, and 3 spring rolls. 8 + 5 + 3 = 16. Rita bought 16 items altogether.

22. Answer: 5 < 7
Explanation: There are 5 broccoli and 7 cabbages. Comparing 5 is less than 7. The answer is 5 < 7.

23. Answer: 85 > 59
Explanation:
8 tens 5 ones = 80 + 5 = 85,
5 tens 9 ones = 50 + 9 = 59.
85 is greater than 59,
The answer is 85 > 59.

24. Answer: 19 drawings
Explanation: Total drawings in the book: 29; Already completed drawings: 10; 29 − 10 = 19 Lisa has 19 drawings to do.

25. Answer: B
Explanation: Eighty-three comes after eighty-two.

26. Answer: A
Explanation: 4 tens and 5 ones = 40 + 5 = 45
98 − 45 = 53; 53 means 5 tens and 3 ones.

27. Answer: 9 tens and 3 ones.
Explanation: 93 is 9 tens and 3 ones.

28. Answer: D
Explanation: There are 30 flowers on the left side and 10 flowers on the right side.
30 + 10 = 40. 40 means 4 tens and 0 ones.

29. Answer: B
Explanation: Given 50 cubes; She gave 2 tens to her friend. 2 tens = 20; 50 − 20 = 30 cubes. Ellen has 30 cubes.

30. Answer: C
Explanation: 40 cubes on the left and 23 cubes on the right side. 40 > 23.

31. Answer: 7 meters is the longest one
Explanation: Given that, 2 shelves are 4 meters and 7 meters long. Upon comparing the numbers 4 and 7; 7 meters is the longest one.

32. Answer: A
Explanation: The digital clock shows Nine thirty.

33. Answer: Missing numbers are 48 and 78.
Explanation:
Here 58 − 10 = 48 and 68 + 10 = 78
So, the missing numbers are 48 and 78.

34. Answer: Blue rope, black rope, and red rope
Explanation: A red rope is 12 cm, a blue rope 30 cm, and a black rope is 22 cm.

35. Answer: 3 items
Explanation: 3 items are not shaded.

36. Answers: D
Explanation: Noah saw 9 gold fishes

ANSWERS AND EXPLANATIONS

37. Answers: A
Explanation: Heena bought three different items, which are avocados, apples, and pizza. The number of avocados he bought is 7, the number of apples he bought is 5 and the number of pizzas he bought is 4. Out of the three items, the one he bought the most was the avocados.

38. Answers: B
Explanation: There are 8 cars, 10 trains, and 6 airplanes. Therefore, the number of airplanes is the fewest.

39. Answers: A
Explanation: According to the tally numbers note that, the number of strawberries is 5, the number of cookies and creams is 4, and the number of fudges is 3. Out of the three it can be noted that strawberries are the most preferred one.

40. Answers: A
Explanation: The sum of the number of stickers collected by Steffi and Angel is 9 + 3 = 12

41. Answers: D
Explanation: From the bar graph given, note that there are 3 different categories of vegetables, which are broccoli, carrots, and cabbage. It can be noted that the number of students who like Broccoli is 8.

42. Answers: B
Explanation: From the given chart, note that, there are three different vegetables, which are cucumbers, beets, and broccoli. Also, note that 4 people like cucumbers, 6 people like beets, and 3 people like broccoli. Therefore, it is clear that beets are the most preferred.

43. Answer: Hexagon
Explanation: Given that, Laura made a shape with 6 edges and corners. Therefore, the required shape will be a hexagon.

44. Answer: B
Explanation: The length of the line is equal to the width of the 4 buttons.

45. Answer: B
Explanation: 28 is 2 ten and 8 ones. James already has 8 ones. He now needs 2 tens.

ANSWERS AND EXPLANATIONS

Made in the USA
Las Vegas, NV
16 October 2023